Energy Conservation in Commercial and Residential Buildings

ENERGY, POWER, AND ENVIRONMENT

A Series of Reference Books and Textbooks

Editor

PHILIP N. POWERS

Professor Emeritus of Nuclear Engineering
Purdue University
West Lafayette, Indiana

Consulting Editor
Energy Management and Conservation
PROFESSOR WILBUR MEIER, JR.

Head, School of Industrial Engineering
Purdue University
West Lafayette, Indiana

1. Heavy Oil Gasification, *edited by Arnold H. Pelofsky*

2. Energy Systems: An Analysis for Engineers and Policy Makers, *edited by James E. Bailey*

3. Nuclear Power, *by James J. Duderstadt*

4. Industrial Energy Conservation, *by Melvin H. Chiogioji*

5. Biomass: Applications, Technology, and Production, *by Nicholas P. Cheremisinoff, Paul N. Cheremisinoff, and Fred Ellerbusch*

6. Solar Energy Technology Handbook: Part A, Engineering Fundamentals; Part B, Applications, Systems Design, and Economics, *edited by William C. Dickenson and Paul N. Cheremisinoff*

7. Solar Voltaic Cells, *by W. D. Johnston, Jr.*

8. Energy Management: Theory and Practice, *by Harold W. Henry, Frederick W. Symonds, Robert A. Bohm, John H. Gibbons, John R. Moore, and William T. Snyder*

9. Utilization of Reject Heat, *edited by Mitchell Olszewski*

10. Waste Energy Utilization Technology, *by Yen-Hsiung Kiang*

11. Coal Handbook, *edited by Robert A. Meyers*

12. Energy Systems in the United States, *by Asad T. Amr, Jack Golden, Robert P. Ouellette, and Paul N. Cheremisinoff*

13. Atmospheric Effects of Waste Heat Discharges, *by Chandrakant M. Bhumralkar and Jill Williams*

14. Energy Conservation in Commercial and Residential Buildings, *by Melvin H. Chiogioji and Eleanor N. Oura*

Additional Volumes in Preparation

Energy Conservation in Commercial and Residential Buildings

MELVIN H. CHIOGIOJI
U.S. Department of Energy
Washington, D.C.

ELEANOR N. OURA
Mele Associates
Silver Spring, Maryland

MARCEL DEKKER, INC.   New York and Basel

Library of Congress Cataloging in Publication Data
Main entry under title:

Energy conservation in commercial and residential
    buildings.

    (Energy, power, and environment ; 14)
    Includes bibliographical references and index.
    Contents: Buildings energy use, 1950-2000 / Eric
Hirst -- Lighting / Robert Dorsey -- Improving appliance
efficiency / Melvin H. Chiogioji -- [etc.]
    1. Buildings--Energy conservation. I. Chiogioji,
Melvin H., [date]. II. Oura, Eleanor N., [date].
III. Series.
TJ163.5.B84E527   690   82-2539
ISBN 0-8247-1874-7   AACR2

MARCEL DEKKER, INC.
270 Madison Avenue, New York, New York 10016

Current printing (last digit):
10 9 8 7 6 5 4 3 2 1

PRINTED IN THE UNITED STATES OF AMERICA

Preface

Buildings account for roughly one-third of total U.S. energy consumption. Energy experts have indicated that we can, by exploiting
currently available technology, cut that energy consumption by 30 to
50% in new buildings and 10 to 30% in existing buildings, with no
significant loss in standard of living, comfort, or convenience.

This book surveys the many architectural/engineering techniques
for combating energy waste in residential and commercial buildings.
Individual chapters discuss overall energy usage in buildings; energy
conservation possibilities in the use of lighting and appliances;
solar heating and cooling; new building design techniques; computer
analysis methodologies; specific techniques for energy conservation
in residential and commercial buildings; the role of the Federal
government in conservation; and implications for the future.

We have asked internationally recognized experts in each of
these areas to expound upon the alternatives available to us regarding both energy conservation and the use of renewable resources in
the residential and commercial buildings sectors. These experts
acquaint us with what is being done and with what can be done in the
design, construction, and maintenance of buildings in order to foster

energy efficiency. Throughout this book, they emphasize life cycle costing as the only sound approach toward energy conservation.

We believe that this text will be a helpful guide to architects and engineers, building owners and developers, contractors, mortgage lenders -- all those who recognize the major challenges that face the buildings industry today.

Melvin H. Chiogioji
Eleanor N. Oura

Introduction

Until recently, many people within the United States considered energy to be both inexhaustible and expendable. These were fairly good assumptions that were verified by the relatively low cost of such energy forms as oil, natural gas, and electricity; and based on these assumptions, the buildings sector of the nation's economy designed and constructed buildings with initial cost as the primary consideration.

Since World War II, our method of building has changed noticeably, as have the materials of building. The mechanical systems have changed, both in response to new comfort standards and to make the buildings built with the new materials and methods tolerable from the Heating, Ventilation and Air Conditioning (HVAC) point of view. Only one factor remains unchanged: energy use per square foot in buildings grows with each succeeding year. We now find that our buildings alone consume about twice the electricity that was used twenty-five years ago.

The large-scale desertion of the city by the middle class and the growth of suburbia have also resulted in substantial energy consumption increases. Suburban living requires a larger per capita

energy use than city dwelling, particularly as each family requires
at least one, and often two or three, cars to get around. Whereas a
radio, a refrigerator, and a range might have been yesterday's list
of energy-consuming appliances, today's inventory includes television sets, washing machines and dryers, tape recorders, freezers,
hi-fi's, dishwashers, air conditioners, and dozens of small appliances that all use energy, generally inefficiently.

Two major public responses have arisen in the wake of a growing
recognition of the energy shortage. The initial response was a call
for the rapid expansion of sources of energy. The other response
is termed "energy conservation." Conserving energy means reducing
the amount of fuel and electricity that a building or space uses
each month. This saves money, and cost savings made now will be
increasingly greater in the future as raw fuel and electricity
prices continue to rise. In addition to fuel and electricity consumption cost savings, other reasons for conserving energy include:

Extending the useful life of existing equipment
Increasing the reserve capacity of existing central plant systems
    and being able to meet future needs without installing extra
    equipment
Reducing the likelihood of shutdown or curtailment of operations
    caused by fuel or power shortages
Reducing air pollution that results from combustion of oil, gas,
    or coal and saving installation costs of pollution control
    equipment

Conservation options made on a cost-effective basis represent a good
investment for the building owner, and conservation can also improve
the quality of housing and the comfort of inhabitants.

It is fairly safe to say that buildings being designed now and
those that will be designed in the future will utilize many of the
new techniques and systems which can lead to maximized energy efficiency. However, because the present building inventory is being
replaced at the rate of only 2 to 3% per year, the majority of existing buildings for many years to come will be those that were not

originally designed with energy conservation in mind. Thus, the
conservation of energy in existing buildings must become a matter
of very real concern as the Nation starts toward a goal of energy
self-sufficiency by the mid-1980s.

The amount of energy consumed in buildings can easily be reduced through the use of technologies which are well understood and
already in use, although in examining the alternative approaches to
conservation, it is important to keep in mind that lowering the rate
of growth in buildings energy use is a behavioral, institutional,
economic, motivational, and, perhaps, structural problem for our
society -- it is not merely a technical problem.

The efficiency of energy use is determined by the three basic
systems that comprise any functioning building: energized systems,
such as those required for heating, cooling, lighting, ventilation,
conveyance, business equipment operation, etc.; nonenergized systems, such as walls, roof, floors, ceilings, windows, etc.; and
human systems, such as maintenance, operation, and management personnel. Each of these systems can be modified for significant savings of energy. However, effective energy conservation requires
that the entire pattern of energy consumption in a building be
analyzed, so that changes made will be integrated into the system
in full light of the interrelationships which exist among the three
systems.

This book brings together the information and design principles
needed to better understand the buildings sector and to design,
construct, or retrofit more energy efficient commercial and residential buildings. Each chapter of the book goes into a different
aspect of buildings energy conservation and improved designs.

In Chapter 1, Eric Hirst discusses long-term trends in residential and commercial energy uses during the second half of the
twentieth century. He provides a number of examples of measures
for reducing energy use in buildings and analyzes alternative futures for residential and commercial energy uses to the year 2000.
He concludes that the range of future energy use in residential and

commercial buildings will depend strongly on both private and government decisions concerning fuel prices, conservation programs, technical improvements, and behavioral changes.

In Chapter 2, Robert T. Dorsey discusses in some detail the effective management of lighting in buildings. Various aspects of lighting and productivity are considered, including the impacts upon
lighting requirements due to age, task difficulty, contrast rendition, and visual comfort probability. A number of lighting conservation techniques are described, and specific suggestions for
implementing these techniques are given. Mr. Dorsey indicates that
optimizing lighting investment involves two basic ideas: (1) buying
the lighting system that gives the most for the dollar invested,
and (2) insuring that that system continues to provide the light
which was originally purchased. This can be done through the use of
the life cycle costing technique.

In Chapter 3, Melvin H. Chiogioji discusses improving appliance
efficiency. Appliances account for about 25 to 30% of the total amount
of energy consumed in the home. Mr. Chiogioji discusses the many
things that can be done to reduce energy consumption in appliances,
both operationally as well as through improved design. The specific
appliances discussed include heating and cooling equipment, hot
water heaters, refrigerators and freezers, ranges and ovens, and
clothes washers and dryers. Appliance energy efficiency standards,
which are projected to improve appliance efficiency by 20 to 30%,
are also discussed.

In Chapter 4, Melvin H. Chiogioji discusses solar heating,
cooling, and hot water systems. Various passive solar systems which
can provide 20 to 95% of the required heat from the sun, without using
active solar collectors, are described. The analysis of existing
systems shows that passive systems are competitive with oil and gas
in some parts of the country. The description of liquid and air
active solar heating, cooling, and hot water systems provides information on the operation of these systems. Methods for sizing
solar heating systems to enable them to provide the desired fraction

of the total heating and/or hot water load of a building are also described, and information on the performance and reliability of existing solar systems is provided. Mr. Chiogioji concludes that the desirability of solar energy use is obvious but that there are technical, legal, and economic difficulties, and even homeowner attitudes, that pose obstacles to more rapid market penetration of solar systems.

In Chapter 5, Charles R. Ince, Jr., discusses buildings design as a national resource. He indicates that when we discuss energy as a design issue, we need to rethink the framework of the energy problem -- i.e., we must substitute the concept of energy consciousness for energy conservation. Whereas energy conservation is identified with systems efficiency in a quantitative approach, energy conscious design implies a qualitative approach to building design that includes (1) integration of energy into initial design decisions concerning program, location, and form; (2) reliance on natural and renewable resources to supply energy; and (3) maximum efficiency in system design and integration. Mr. Ince presents a number of residential and commercial case studies which support his thesis.

In Chapter 6, Maurice Gamze describes and examines eighteen building load modeling computer analysis programs. Each of the eighteen programs were tested using a hypothetical building and using artificial weather conditions. In analyzing these programs, the following phenomena were examined: (1) building thermal storage effects; (2) heat exchange between interior surfaces; and (3) convective and radiative components of heat gains from lights and people. Mr. Gamze concludes that the differences in computed energy requirements are caused by differences in the techniques utilized to model internal heat balances and the building's storage effects and that these differences can result in significant variation in estimating heating and cooling loads. He thus questions the use of these techniques for such broad policy issues as establishing building energy performance standards.

In Chapter 7, Melvin H. Chiogioji discusses energy conservation
in residential buildings and the use of passive design considerations. He stresses that a building can increase energy efficiency
by recognizing and using to advantage the natural elements that surround it and that the building location, siting, orientation, configuration, layout, construction, mechanical and electrical systems,
and interior furnishings should be carefully evaluated in terms of
their contributions to energy consumption and conservation. Energy
savings potential of various improvements to buildings are discussed.
Three case studies which show that energy savings of 50% can be
achieved in buildings designed with energy efficiency in mind are
presented.

In Chapter 8, Kirtland C. Mead discusses energy conservation in
commercial buildings. Past energy usage patterns have indicated
that it is a very fertile area for energy conservation efforts. He
provides substantial statistical profiles on the commercial sector
as well as on its patterns of energy use. The various energy systems in commercial structures are described, and energy conservation
measures are discussed in great detail. Specific areas for assessment include the building envelope, lighting, ventilation and heat
recovery, HVAC systems, and building management and control systems.

In Chapter 9, Maxine L. Savitz describes in detail what she considers to be the federal role in energy conservation. She indicates
that the traditional role which the federal government has taken in
energy matters needs to be modified and that a more aggressive role
is necessary. Dr. Savitz presents six general principles which
should shape the federal government's approach to energy conservation activities and describes the barriers to energy conservation.
Federal government policies for the building sector, among them,
policies in public information, financial incentives, research and
development, and regulations, are discussed in some detail, and
specific recommendations for action are given. Dr. Savitz concludes
that energy conservation is an important element in any national
strategy to combat the rising costs of energy, to reduce our depen-

dence on foreign petroleum sources, and to boost our economic prosperity.

In Chapter 10, John H. Gibbons describes implications for the
future of energy conservation in the buildings sector. He indicates
that our buildings will undoubtedly be supplied with and utilize
energy efficiently in the future and that the shape of the future
will be determined by the alternatives we choose and more importantly,
perhaps, by the rate at which we make the transition from old to
new technologies. If we intend to radically alter energy demand in
the buildings sector in the 1990s, a circumstance that is technically possible and highly desirable in terms of economic and national
security interests, a combination of rising energy prices, strong
regulations, and increased direct federal intervention will be required. He states that the crafting of the national energy policy
will call for new levels of flexibility that will serve both the
national need to conserve energy rapidly and the varying needs of
states, localities, and individuals.

Ultimately, the principal reasons for analyzing and studying
the field of energy use in buildings are to gain a better knowledge
of how to design and redesign buildings and to apply this knowledge
to our built environment. In the future, our buildings must have
less reverence for and adherence to symmetry and geometric formalism.
The concept of the building as a static ideal object in space will
yield to the idea that it is a part of a growing, changing, continuous process.

The reexamination of existing buildings for continued usage or
for conversion to new uses should assume a greater importance. The
amount of energy required for the alteration of older buildings,
for example, is substantially less than the amount required for
building their replacements, and often the older buildings, designed
to be comfortable with less dependence on mechanical systems, are
able to function in the future with the same economy of energy consumption.

Rather than becoming less important, the city should become
more important. Its future development will result in cities which

have less density at their cores, and will be extensively inter-
laced with green areas. These green areas can serve not only as
recreational relief to the built areas, but can also offset the
heat retention rate of the buildings and streets, serve as utility
distribution channels, house decentralized total energy plants serving these smaller communities, and may act to systematize the mass
transportation network that must replace individual vehicles.

Power and heat should be provided through a varied combination
of sources. The sun can do more than heat the transfer device--
the solar collectors; it can be used more directly to heat building
walls and interiors and also to bring air movement and light to the
spaces that need them. There will also be a rapid growth of smaller
decentralized electrical generating units close enough to their load
centers to make efficient use of what would otherwise be waste heat.

As these changes are undertaken, we will see the shape of the
future. Evolving from the building forms of today will be the more
humanized, more climate responsive, more visually intricate, and
more nature oriented world of tomorrow.

Let us see how this can be achieved.

Contributors

MELVIN H. CHIOGIOJI Deputy Assistant Secretary for State and Local
  Assistance Programs, U.S. Department of Energy, Washington, D.C.

ROBERT T. DORSEY* Manager, Lighting Technology Development, Lamp
  Marketing and Sales Operation, General Electric Lighting Business Group, Nela Park, Cleveland, Ohio

MAURICE GAMZE President, Gamze-Korobkin-Caloger, Inc., Chicago,
  Illinois

JOHN H. GIBBONS Director, Office of Technology Assessment, U.S.
  Congress, Washington, D.C.

ERIC HIRST Research Engineer, Oak Ridge National Laboratory, Oak
  Ridge, Tennessee

CHARLES R. INCE, JR. President, American Institute of Architects
  Research Corporation, Washington, D.C.

KIRTLAND C. MEAD Principal, Resource Planning Associates, Inc.,
  Paris, France, and Cambridge, Massachusetts

MAXINE L. SAVITZ Deputy Assistant Secretary for Conservation, U.S.
  Department of Energy, Washington, D.C.

---

*Present affiliation: Professional Engineer, 3726 Tolland Road,
Shaker Heights, Ohio

Contents

PREFACE   *iii*

INTRODUCTION   *v*

CONTRIBUTORS   *xiii*

1. BUILDINGS ENERGY USE: 1950-2000   *1*
    Eric Hirst

    History   *2*
    Measures to Reduce Energy Use   *11*
    Barriers to Conservation   *22*
    Future Energy Use   *26*
    Conclusion   *36*
    Notes and References   *38*

2. LIGHTING   *41*
    Robert T. Dorsey

    Putting Lighting in Perspective   *42*
    Lighting and Productivity   *43*
    Energy Conservation Practices   *52*
    Light Sources   *57*
    Nonuniform Illumination   *64*
    Lighting Controls   *65*
    Luminaires   *70*
    Ballasts   *77*
    Daylighting   *79*
    Total Building Thermal Operation   *83*
    Conclusion   *84*
    Notes and References   *87*

3. IMPROVING APPLIANCE EFFICIENCY   *89*
    Melvin H. Chiogioji

    Heating and Cooling Equipment   *91*
    Hot Water Heaters   *108*
    Refrigerators and Freezers   *117*
    Ranges and Ovens   *126*
    Clothes Washers and Dryers   *131*
    Integrated Appliances   *132*
    Appliance Efficiency Standards   *133*
    Conclusion   *136*
    Notes and References   *137*

4. SOLAR HEATING/COOLING/HOT WATER   *139*
    Melvin H. Chiogioji

    Passive Solar Design   *140*
    Active Solar Systems   *146*
    Solar Collectors   *149*
    Thermal Storage   *155*
    Solar Space Heating Systems   *158*
    Solar Water Heating   *164*
    Solar Cooling Systems   *171*
    Solar-Assisted Heat Pumps   *175*
    Solar System Sizing   *178*
    Performance of Solar Heating and Cooling Systems   *184*
    Conclusion   *189*
    Notes and References   *195*

5. DESIGN, OUR VISIBLE NATIONAL RESOURCE   *197*
    Charles R. Ince, Jr.

    Definition of Terms   *204*
    Energy Performance Standards   *213*
    The Practice of Design: Case Studies   *217*
    The Shape of the Future   *232*
    Conclusion   *234*
    Notes and References   *235*

6. BUILDING COMPUTER ANALYSIS METHODOLOGIES   *237*
    Maurice Gamze

    Load Modeling   *239*
    Program Analyses   *244*
    Conclusion   *267*
    Notes and References   *269*

7. ENERGY CONSERVATION IN RESIDENTIAL BUILDINGS   *271*
    Melvin H. Chiogioji

    Building Siting and Orientation   *274*
    Building Envelopes   *285*
    Energy Savings Potential   *303*
    Case Studies   *309*
    Conclusion   *329*
    Notes and References   *329*

8. ENERGY CONSERVATION IN COMMERCIAL BUILDINGS   *331*
   Kirtland C. Mead

   The Commercial Sector   *333*
   The Office Building and Its Energy Systems   *337*
   Thermal Performance of the Building Shell   *349*
   Lighting   *361*
   Ventilation and Heat Recovery   *365*
   HVAC Systems   *372*
   Central HVAC Plant   *376*
   Operations and Maintenance   *378*
   Building Management and Control   *379*
   Building Analysis Techniques   *382*
   Notes and References   *384*

9. THE ROLE OF THE FEDERAL GOVERNMENT   *385*
   Maxine L. Savitz

   The Nature of Our Current Building Stock   *387*
   The Traditional Federal Role in Energy   *389*
   The Federal Role Today   *390*
   Conclusion   *408*
   Notes and References   *409*

10. IMPLICATIONS FOR THE FUTURE   *411*
    John H. Gibbons

    Energy Cost and Availability   *413*
    Energy Efficiency and Resilience   *416*
    Energy Savings and Consumer Costs   *417*
    Opportunities Through Research   *418*
    Economics   *422*
    A RD&D Agenda for Conservation   *423*
    Making the Transition   *426*
    Notes and References   *431*

LIST OF FIGURES   *433*

LIST OF TABLES   *437*

INDEX   *441*

1

Buildings Energy Use: 1950-2000

Eric Hirst

Oak Ridge National Laboratory
Oak Ridge, Tennessee

The purpose of this chapter is to discuss long-term trends in residential and commercial (R/C) energy uses during the second half of the twentieth century. The topics covered include: definitions of the residential and commercial sectors; historical patterns of energy use in the buildings sectors (by fuel, end use, and region); types and examples of measures for reducing energy use in buildings; and alternative futures for residential and commercial energy uses to the year 2000.

The buildings sectors account for one-third of the nation's energy budget: 28 QBtu (1) in 1978, the cost of which was about $80 billion that year (2). Because of rapidly rising fuel prices during the early and mid-1970s, annual fuel bills to owners and occupants of residential and commercial buildings have been climbing steeply. To partly offset this economic burden of rising fuel prices, various energy conservation measures can be put into effect. One kind of conservation measure involves operational changes, such as reducing winter thermostat settings. A second kind involves improvements in the technical efficiencies of the systems that heat,

cool, and light buildings, such as installing additional insulation
on heating/cooling ducts. Both of these measures are discussed
later in the chapter.

The potential energy and monetary benefits of these kinds of
conservation measures are surprisingly large; however, achieving
these potential benefits is far from easy because of a variety of
personal, social, institutional, and political barriers to conservation. A simple example is the family that does not reduce its winter thermostat setting because it is unaware of the large dollar
savings this measure can yield. Thus, lack of adequate site-
specific information is probably an important barrier to adoption
of cost-effective conservation measures.

In order to explore the implications of different kinds of
government programs designed to overcome these barriers, alternative
projections of buildings energy use up to the end of the century
have been developed. These projections show the large national
energy and economic benefits possible with adoption of conservation
programs. In addition, some of the more important second-order effects of these programs (i.e., the effects on employment, foreign
trade, and net energy consumption) are discussed.

HISTORY

The residential sector is defined as those structures (single-
family units, multifamily units, mobile homes) occupied by households (both families and unrelated individuals). Group quarters,
such as jails, hotels, and hospitals, are considered part of the
commercial sector, which is defined as those structures (e.g.,
office buildings, schools, hospitals, stores) that house the service sectors of our economy (e.g., retail and wholesale trade,
finance and insurance, government enterprises).

The number of households in the United Stares increased from
44 million in 1950 to 71 million in 1975 and 76 million in 1978
(see Table 1-1). The average growth rate in household formation

Table 1-1. The Residential Sector: 1950-1978

	Population (millions)	House holds (millions)	Distribution by Housing Type (%)		
			Single-Family	Multi-Family	Mobile Home
1950	152	43.6	66	33	1
1960	180	52.8	75	24	1
1970	204	63.4	69	28	3
1975	213	71.1	68	28	4
1976	215	72.9	68	28	5
1977	216	74.1	68	27	5
1978	218	76.0	—	—	—

Source: Refs. 3 and 4.

during this 28-year period was 2.0% per year, while the population growth rate was 1.3% per year (3). Thus, average household size dropped from almost 3.4 persons in 1950 to less than 2.9 in 1978. Much of this decline is due to the increase in the number of unmarried adults ("primary individuals," as defined by the Bureau of the Census) who form their own households, undoubtedly a reflection of growing income.

Declining household size (i.e., more rapid growth in number of households than in population) has two effects with respect to residential energy use. The increase in households relative to population leads to greater energy use. Partly offsetting this is the reduction in per household energy use because of fewer members (e.g., fewer showers, less laundry).

During this period, single-family homes were the dominant housing choice for Americans, accounting for just over two-thirds of the total occupied housing stock in both 1950 and 1977 (4). Multifamily units (both low-rise and high-rise structures) account for most of the remaining housing units. During the past decade mobile homes have become increasingly popular, expanding their share of the total occupied housing stock from 1% in 1960 to 3% in 1970 and 5% in 1977 (Table 1-1).

Households in single-family units tend to use much more energy, particularly for heating and air conditioning, than do households occupying the other types of housing units (see Table 1-2). On a per-unit-area basis, multifamily units are the most efficient, because they have fewer surfaces exposed to the outside than either of the other two housing types. A typical unit in a large apartment house may have only one exterior wall; the other walls and the floor and ceiling may all be facing other apartments. On the other hand, mobile homes are relatively inefficient because of their poor geometric shape (long and narrow, which leads to a high ratio of surface area to volume) and inferior quality of construction.

Four building types dominate the commercial sector. In 1975, offices accounted for 15% of total floorspace, retail/wholesale for 19%, hospitals for 7%, and educational buildings for 23%. Combined, these four building types account for almost two-thirds of total commercial floorspace (5).

Table 1-3 shows the growth of commercial floorspace from 1950 to 1978 and the distribution of that floorspace among the major building types (6). Between 1950 and 1978, growth in commercial floorspace averaged 3.7% per year, while growth in the gross national product was slower, at 3.5% per year. This higher growth in floorspace is a reflection of the changing composition of the GNP: A steady shift to a services economy. During the past few years of this period, growth in both floorspace and GNP was much slower than it had been during the 1950s and 1960s.

Table 1-2. Approximate Differences in Residential Space Heating Energy Use by Housing Type

	Typical Size (ft^2)	Relative Energy Use*	
		(per unit)	(per ft^2)
Single-Family	1,500	1.0	1.0
Multifamily	1,000	0.4	0.6
Mobile Home	800	0.6	1.2

*Refers to space heating energy use/space heating energy use for single-family unit.

Table 1-3. The Commercial Sector: 1950-1978

	Floorspace (billion ft²)					GNP (trillion 1975-$)
	"Commercial"*	Educational	Public	Hospital	Total**	
1950	4.7	2.7	0.5	0.9	11.1	0.68
1960	6.3	4.2	0.7	1.2	15.8	0.94
1970	9.8	6.0	1.0	1.7	24.3	1.37
1975	12.3	6.6	1.2	2.0	28.3	1.52
1976***	12.8	6.7	1.2	2.1	29.2	1.62
1977***	13.5	6.7	1.2	2.1	30.3	1.70
1978***	—	—	—	—	30.6	1.76

*Commercial includes office, retail/wholesale, garage, and warehouse. In 1975, for example, offices accounted for 44% of the commercial subtotal, retail/wholesale for 34%, garage for 5%, and warehouse for 18%.

**Total includes religious, hotel/motal, and miscellaneous buildings not detailed elsewhere.

***Floorspace figures for 1976, 1977, and 1978 are estimates developed at ORNL.

Source: Refs. 3, 5, 6 and Bureau of Economic Analysis, *Survey Curr. Bus.,* vol. 58, no. 3, March 1978.

Energy use in the combined buildings sector grew from 9.6 QBtu in 1950 to 28.3 QBtu in 1978, with an average annual growth rate of 3.9% per year (7). As Fig. 1-1 shows, growth was rapid and steady from 1950 to 1973, with an average growth rate of 4.2% per year; however, since 1973, growth has been slow and erratic (2.6% per year).

The importance of coal during this 28-year period changed dramatically: in 1950 direct use of coal accounted for 30% of the sector's fuel use, while in 1978 it accounted for only 1% of the total. Oil's share dropped from 27 to 20% during this period. On the other hand, shares accounted for by electricity and gas increased from 25 and 17% to 49 and 30%, respectively.

Figure 1-2 shows the distribution of fuel between the residential and commercial sectors (7). In 1950, the residential sector accounted for almost 70% of the R/C sector's fuel use, while in 1978 the residential sector accounted for only about 60%. Thus, while energy use in households grew at the rate of 3.6% per year, energy use in commercial buildings grew at the rate of 4.5% per year.

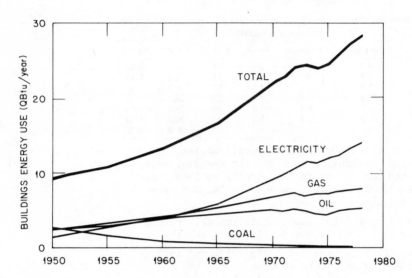

Figure 1-1. Buildings energy use by fuel type; 1950-1978.

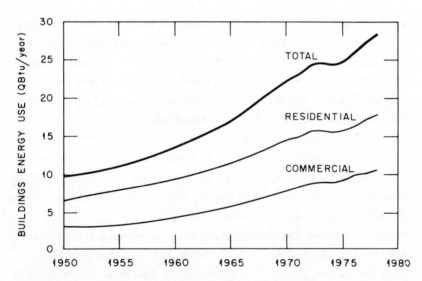

Figure 1-2.   Energy use in residential and commercial buildings;
1950-1978.

These historical differences are likely to continue in the future as
the service sectors of our economy grow at a faster rate than the
overall GNP.

Figure 1-3 shows trends in residential fuel prices for the same
period (8). Generally, prices were declining or stable, until the
early 1970s: Since then prices for all fuels, especially for gas
and oil, have risen. Real oil prices increased 65% between 1972
and 1978, while gas prices increased 35% and electricity prices rose
15%. Trends in commercial sector fuel prices are similar to those
shown for residential prices in Fig. 1-3.

Tables 1-4 and 1-5 detail energy use by fuel/end use combination
in residential and commercial buildings (9). The most important end
use is space heating, which accounts for 52% of residential energy

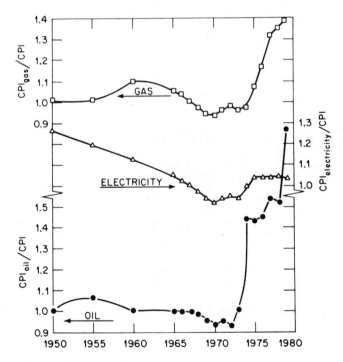

*Figure 1-3. Trends in real fuel prices to the residential sector,
1950-1978. (From Ref. 8.)*

Table 1-4. *Residential Energy Use by Fuel/End Use,*
1975

	Electricity	Gas	Oil	Other*	Total
		(QBtu)			
Space Heating	1.36	3.81	2.35	0.54	8.06
Water Heating	1.05	0.96	0.18	0.05	2.24
Refrigerators	0.92	—	—	—	0.92
Freezers	0.38	—	—	—	0.38
Cooking	0.46	0.29	—	0.01	0.76
Air Conditioning	1.08	—	—	—	1.08
Lighting	0.90	—	—	—	0.90
Other	0.86	0.45	—	—	1.31
Total	7.01	5.51	2.53	0.60	15.65

*Other fuels include coal, coke, and LPG.

Source: Ref. 9.

Table 1-5. *Commercial Energy Use by Fuel/End Use,*
1975

	Electricity	Gas	Oil	Other*	Total
		(QBtu)			
Space Heating	0.33	1.66	1.88	0.12	3.99
Air Conditioning	1.83	0.14	—	—	1.97
Water Heating	0.04	0.08	0.10	—	0.22
Lighting	2.09	—	—	—	2.09
Other	0.76	0.17	—	—	0.93
Total	5.05	2.05	1.98	0.12	9.20

*Other fuels include coal, coke, and LPG.

Source: Ref. 9.

use and 43% of commercial energy use. Water heating is the second most important fuel in homes (14% of the total used) but is much less significant in commercial buildings (2). Lighting and air conditioning, on the other hand, are much more important energy uses in commercial buildings than in residential buildings. The numbers in Tables 1-4 and 1-5 suggest that the major opportunities for energy conservation are in space heating and cooling, water heating in homes, and lighting in commercial buildings.

Tables 1-4 and 1-5 also show that the contribution to different end uses varies across the fuels. Fossil fuels (gas, oil, liquid petroleum gas; LPG) are used almost exclusively for heating and water heating. Electricity is used for a variety of purposes; only 25% of electricity is for heating and water heating.

Table 1-6 and Figs. 1-4 and 1-5 show the distribution of residential and commercial energy uses across the major regions of the United States (10,11). Table 1-6 also shows the population, number

Table 1-6. *Residential and Commercial Energy Use by Federal Region, 1975*

Region		Population (millions)	Households (millions)	Commercial Floorspace (billion ft^2)	Energy Use (QBtu)	
					Residential	Commercial
1	New Eng	12.2	4.1	1.61	0.86	0.69
2	NY/NJ	25.6	8.5	3.88	1.67	1.21
3	Mid-Atl	24.2	8.1	3.17	1.66	0.93
4	S-Atl	34.9	11.6	4.19	2.40	1.21
5	Midwest	45.2	15.1	5.87	3.91	1.92
6	S-West	21.9	7.3	2.77	1.57	0.95
7	Central	11.5	3.8	1.49	1.01	0.51
8	N-Central	6.2	2.1	0.79	0.48	0.35
9	West	24.9	8.3	3.63	1.48	1.05
10	N-West	7.0	2.3	0.91	0.63	0.37
U.S.		213.6	71.2	28.31	15.67	9.20

Source: Refs. 10 and 11.

of households, and amount of commercial floorspace for each of these
10 federal regions. Much of the difference in energy use between
the regions results from variations in the number of households (for
residential energy use) or the stock of floorspace (for commercial
energy use). The maps in Figs. 1-4 and 1-5 show the variation in
per unit energy use (Btu's per household and Btu's per square foot
of commercial floorspace). Climate, fuel prices, and historical fuel
choices also influence energy use patterns. For example, energy use
is highest in the Northwest (270 million Btu per household and
410,000 Btu/ft^2) because of its cold climate and very low electricity
prices (due to the presence of the Bonneville Power Administration).
Energy use is lowest in the West (180 million Btu per household and
290,000 Btu/ft^2) because of its very mild climate and slightly higher
than average fuel prices.

The numbers presented in this section show the diversity in
residential and commercial buildings, and of energy use patterns by
type of fuel, region, and end use over time (from 1950 to 1977).
These data and estimates suggest that buildings energy use is a complicated function of demographics (population, age distribution of
the population, regional migration, households), of economic activity,

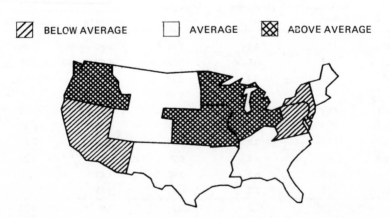

*Figure 1-4. Regional variation in 1975 residential energy use per
household. Regions labeled "average" used between 200 and 250 million Btu/household; the national average was 220 million Btu/house-
hold in 1975.*

climate, and historical practices. We can briefly summarize the
historical data presented by referring again to Figs. 1-1 to 1-3.
Energy use grew steadily and rapidly from 1950 to the early 1970s.
Since then, growth has been slow and erratic. This change in energy
use trends is closely paralleled by changes in real fuel prices,
shown in Fig. 1-3. When fuel prices began to rise sharply in the
early and mid-1970s, energy growth faltered. A major question
arises: What will future trends in fuel prices and in buildings
energy use be? The rest of this chapter is devoted to the second
part of this question, but we must first consider several specific
measures that can be implemented to reduce energy use.

MEASURES TO REDUCE ENERGY USE

Basically, there are two ways to cut energy use in buildings. The
first involves changes in the way we operate existing systems. Such
changes include the raising of thermostat settings on room air conditioners; the lowering of hot water temperature settings; better

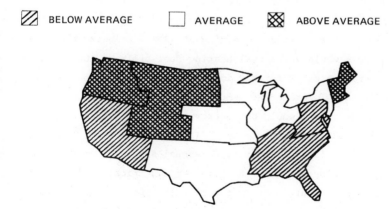

*Figure 1-5. 1975 Commercial energy use per square foot. Regional
variation in 1975 commercial energy use per square foot of floorspace. Regions labeled "average" used between 300,000 and 400,000
Btu/ft^2; the U.S. national average was 320,000 Btu/ft^2 in 1975.*

control of office building heating, ventilation, and air conditioning (HVAC) systems during hours of nonuse; and a reduction in the
amount of lighting in commercial buildings. These changes are characterized by their low (or zero) capital cost, the speed with which
they can be implemented, and the fact that they all require some
kind of behavioral change.

The second class of conservation measures involves improvement
in the efficiency of the equipment used in buildings and of the
structures themselves. Examples include the addition of insulation
in attics and under floors, the replacement of electric resistance
heating systems with heat pumps, the use of fluorescent rather than
incandescent lighting, and the elimination of pilot lights on gas-
burning equipment. These technological changes are characterized by
minimal behavioral changes, some (occasionally high) capital cost,
and delays in implementation (as existing equipment wears out).

Space heating is the dominant energy use in homes; therefore,
specific conservation measures should begin with a consideration of
the way in which heating systems operate. The greatest energy savings can be gained by reducing temperature setting during the winter
(see Fig. 1-6) (12). Consider as an example a single-family home
located in Kansas City (with about 5000 heating degree-days, HDD)
which keeps a winter thermostat setting at 72°F. Reducing the temperature to 70°F would cut annual heating fuel bills by 10%. Lowering the temperature still further to 68°F would increase savings to
20%. In addition, lowering the temperature at night (10 p.m. to
6 a.m.) from 68 to 60°F would cut energy use in the Kansas City home
by a remarkable 30% (relative to the 72°F reference point). This
30% energy savings is equivalent to an annual saving of about $100
for a gas-heated home and about $250 for an electrically heated home.
No investments are required to implement this type of conservation
measure except, perhaps, a clock thermostat ($75 to $100) and additional sweaters for the family.

As Fig. 1-6 shows, the energy benefits of winter temperature
reductions vary from region to region, depending on the severity of
the winter (as measured by HDD). In northern climates, the *percentage*

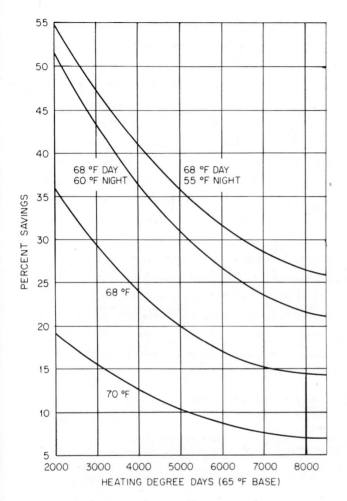

Figure 1-6. Predicted energy savings for several thermostat set-tings. (From Ref. 12.)

reduction in heating energy use is relatively small; however, the *absolute* value of the energy (and dollar) savings is larger than in mild southern climates. The same kinds of energy and dollar savings are possible by setting temperatures *up* in the summer to reduce air conditioner operating time. As a rough rule of thumb, a 1°F increase in temperature setting will reduce air conditioner use by about 5% (13). Thus, raising the thermostat from 72 to 78°F would cut air conditioner electricity use by more than 25% and, depending on local electricity prices, climate, and household habits, the annual dollar savings would be about $15.

The use of natural ventilation can also cut summer air conditioning requirements. Many homes with central air conditioning keep the windows closed all summer long; however, there are often times during the summer when the outdoor air temperature is low enough for natural ventilation to eliminate the need for air conditioning. Table 1-7 shows both the number of hours an air conditioner compressor operates in nonventilated houses in different cities and the percentage reduction in compressor operation when natural ventilation is used (14). The percentage reduction in air conditioner use (and therefore in air conditioning electricity consumption) ranges from a low of 12% in Phoenix to a high of 73% in San Diego. Roughly speaking, one-half of total air conditioning energy use could be eliminated if natural ventiliation were used, because there are many hours with outside summer temperatures low enough to offset internal heat generation resulting from appliances and people in the home.

Table 1-7. Effects of Natural Ventilation on Residential Air Conditioning Requirements

	Annual Compressor Operating Hours in an Unventilated House	Percent Reduction in Operating Hours Due to Use of Natural Ventilation
Miami	2900	25
New Orleans	2310	18
Phoenix	2120	12
Dallas	2000	20
Atlanta	1520	35
Topeka	930	33
New York	760	48
Chicago	730	37
Minneapolis	590	37
San Diego	590	73

Source: Ref. 14.

In a similar fashion, operational changes can cut energy use substantially in commercial buildings. A 40,500-ft^2 office building in Kansas City serves as an example (see Fig. 1-7). Baseline energy use for the heating, cooling, and ventilation of the preembargo design building is about 430,000 Btu/ft^2-year (15). Changing thermostat settings on weekdays between 6 p.m. and 6 a.m. and for the weekend can cut annual energy use by 37%. Reducing lighting levels and requiring natural ventilation can save another 12%. Finally, using zone demand reset for deck temperatures can save another 9%. Altogether, these operational changes can cut energy use in this typical office building by more than half, reducing it from 430,000 to 180,000 Btu/ft^2-year.

This estimated reduction in office building energy use was borne out in Los Angeles during the 1973-1974 Arab oil embargo. Because much of Los Angeles' electricity supply is oil-generated, the need for immediate curtailment measures was important. The commercial sector was able to cut its electricity consumption by 20 to 30% relative to consumption for the same months during the previous year (16). Much of this savings was attributed to reductions in lighting levels, and a combination of changes, including fewer hours of operation for HVAC systems and higher temperature settings for air conditioning systems, accounted for the rest. Interviews with building managers suggested that these conservation measures were simple to implement and that the economic benefits gained were sufficient to ensure continuation of the measures after the embargo.

The second broad class of conservation measures encompasses improvements in the technical efficiencies of equipment; appliances, and structures. We can subdivide this approach into two aspects -- one deals with modifications to systems already in use (retrofit) and the other deals with the purchase of new appliances, equipment, and structures.

Turning once again to the Kansas City home discussed earlier, we consider retrofit measures as they apply to residential space heating. Assuming that the home was built in the late

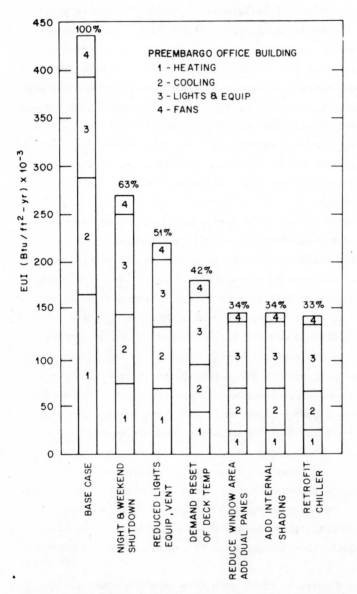

Figure 1-7. Effects of energy conservation practices and measures on annual energy use in an office building constructed during the 1960s.   (From Ref. 15.)

1960s and that it met federal standards then in effect, this house
would have 3 in. of insulation (R-11) in the attic, none in the walls
or floor, and no storm windows or doors. Adding 3 in. (R-11) to the
floor, 6 in. (R-19) to the attic, and installing storm windows would
cut its annual heating bills by about one-fourth. Annual savings
for a gas-heated house would be about $75; for an electrically heated
house savings would be about $200. Since the total cost of this
retrofit package would be about $750, it would take 10 years to re-
cover this investment in a gas-heated home and less than 4 years in
an electrically heated one.

Let us now consider a *new* house, again in Kansas City. Because
it is a new home rather than an existing one, insulation would be
used in the exterior wall cavities [3½ in. (R-13)], in the attic
(R-30), and in the floor (R-11). Orienting the house on the building site so that most of its windows are on the long south face would
ensure that the sun's energy entering the house in winter provides
free energy, often called "passive solar." In the summer, the large
overhang on the south face would block the sun's rays, so that no
additional load would be imposed on the air conditioner. As in the
retrofit house, storm windows would be installed. Finally, the
heating system would be properly sized, since traditionally typical
heating systems have twice the capacity needed.

These measures would reduce the heating load by over 40% and
the air conditioning load by 15%, relative to the original
design (see Figs. 1-8 and 1-9) (17). These reductions mean that
smaller heating and cooling equipment can be installed in the new
house, providing a dollar savings that partially offsets the higher
cost of the improved structure. Thus, the new house costs only
about $550 more than the existing house. For a gas-heated home, the
annual fuel bill can be cut by $130; for an electrically heated
home, the annual dollar savings is $330.

Efficiency improvement options are also available for heating
and cooling equipment and for major household appliances such as
water heaters and refrigerators. Figures 1-10 and 1-11, which cover

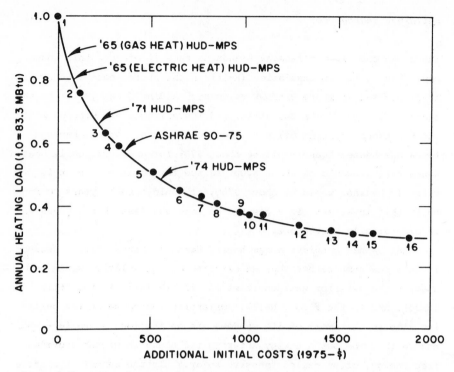

Figure 1-8. Annual space heating load for a new Kansas City single-family home versus additional initial cost. (From Ref. 17.)

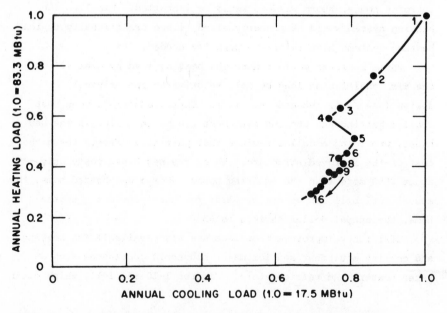

Figure 1-9. Relationships between annual heating and cooling loads for a new Kansas City single-family home. (From Ref. 17.)

annual energy use of residential gas furnaces (18) and refrigerators
(19), are similar in shape to the curves shown for structures (Figs.
1-8 and 1-9). A combination of several design changes to a new gas
furnace, such as those shown in Fig. 1-10, could cut space heating
energy use by at least 25%. The cost to the homeowner for the furnace would increase by 22% ($300); however, the reduction in annual
fuel bills would be sufficient to repay this investment in 6 years.
Because the furnace is likely to last at least 15 years and because
the price of gas is almost certain to increase in the future, this
is an attractive investment. The potential savings in refrigerators
is even larger, as shown in Fig. 1-11. Energy use could be reduced
by as much as 50% with a pay-back period of less than 3 years.

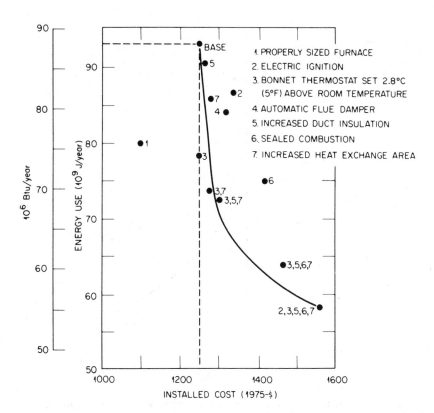

Figure 1-10. Annual energy use versus initial cost for a typical
gas furnace. (From Ref. 18.)

The same kinds of opportunities to reduce energy consumption
exist with commercial buildings and their HVAC systems. Figure 1-12
shows the relationship between annual energy use and capital cost
for HVAC systems in the same Kansas City office building considered
earlier (20). Two curves are shown: the dashed curve shows the relationship between HVAC energy use and capital cost for systems installed in new buildings; the solid curve refers to modifications
to existing systems in existing buildings. Clearly, it is less

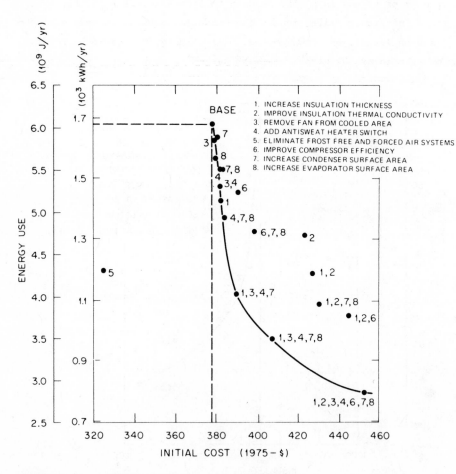

*Figure 1-11. Annual energy use versus initial cost for a typical
refrigerator. (From Ref. 19.)*

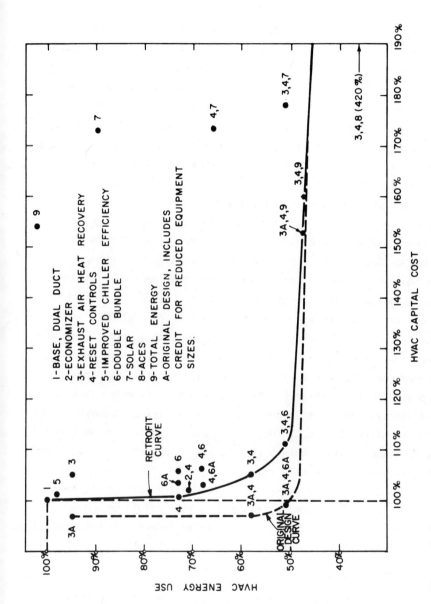

Figure 1-12. Engineering analysis of HVAC systems. (From Ref. 20.)

expensive to install energy-saving equipment when designing a new
building than to modify an existing one.

Some of the design changes save remarkable amounts of energy
and have very short pay-back periods. For example, the combined use
of reset controls (item 4 in Fig. 1-12) and double-bundle heat exchangers (item 6) can cut HVAC energy consumption by more than 30%.
The pay-back period in a new office building is 1 year (2 years in
an existing building). These figures also suggest that reductions
beyond 50% in HVAC energy use become progressively more expensive
(i.e., the curve becomes very flat), leading to long payback periods
at today's fuel prices.

The conservation measures discussed in this section are all
available today. They do not require further research before they
are ready for commercial applications. This array of present-day
techniques for cutting energy use is impressive, but perhaps even
more impressive are the economic benefits that generally accompany
implementation of these measures. In addition, both private industry and the federal government (Department of Energy; DOE) are spon-
soring programs to develop new technologies such as electrodeless
fluorescent lamps, gas-fired heat pumps, integrated appliances, and
improved control systems, all of which will provide buildings functions with smaller energy inputs. For example, research to develop
and field-test an electric-heat-pump water heater is underway. The
system consists of a small compressor mounted above a conventional
electric water heater storage tank (21). Laboratory tests suggest
that this water heater will require only half as much electricity
as a conventional water heater does. The extra cost of the heat
pump water heater is expected to be less than $200. Thus the extra
investment will be repaid within 2 years at today's electricity prices.

BARRIERS TO CONSERVATION

The variety and cost-effectiveness of the conservation measures discussed in the preceding section make one wonder whether they are
being adopted by owners and occupants of residential and commercial

buildings. Data on recent trends in both energy use and conservation actions are extremely hard to obtain and even more difficult to interpret. However, it appears that in the buildings sectors, improvements in energy efficiency are occurring, albeit slowly. Several plausible reasons exist for this slow pace.

First, we are influenced by a long historical record of low and declining fuel prices (Fig. 1-3). Indeed, until the early 1970s, the energy cost of operating an HVAC system was a small--and declining--fraction of its total cost (purchase, maintenance, operation). Therefore, little attention was devoted to the energy efficiency of the system. Now, with fuel prices suddenly much higher and likely to rise even more, we need to change the way we think about, design, purchase, and operate these systems. However, inertia is a common characteristic of people and their institutions: it will take time to adjust to higher fuel prices.

The need for information and the difficulty of effectively processing that information also contributes to the slow pace of increasing energy efficiency. Adoption of cost-effective conservation measures can only take place when decision-makers (homeowners, tenants, etc.) have accurate, site-specific information concerning the costs and benefits of their options. Such information is presently not readily available to building owners and operators, although many programs are underway to remedy that deficiency. Several states have implemented energy "hotlines" which provide telephone answers to frequently asked questions. In addition, many electric and gas utility companies offer residential and commercial customers on-site audits of their buildings. This service is now required for residential customers under the National Energy Act (NEA; National Energy Conservation Policy Act of 1978).

Information is a necessary prerequisite to adoption of wise conservation measures; however, information by itself is not sufficient. For example, most technical efficiency improvements to equipment and structures involve higher initial costs, a fact which decision-makers must learn to deal with. Offsetting this higher first

cost is the subsequent reduction in operating energy costs. The
balancing of initial versus operation costs can be effectively handled with an approach called *life-cycle costing,* the life-cycle cost
of a particular system being the sum of the initial cost plus the
present value of its lifetime operating cost. The *present value* approach allows one to incorporate the time value of money (for
example, home mortgage loans which generally have interests rates
of 10 to 20% per year). In principle, the system with the lowest
life-cycle cost should be chosen, but purchase decisions have traditionally been based on initial cost. As more and more decision-
makers recognize the upward trend in fuel prices, the notion of
life-cycle costing is likely to become the foundation of energy-
related purchase decisions.

A third obstacle to speedy adoption of conservation measures is
the fragmented nature of the buildings sector. The design, construction, financing, and operation of residential and commercial buildings involve a multitude of participants: homeowners and renters,
commercial firms, developers, architects, engineers, builders, contractors, labor unions, financial institutions, insurance firms, and
government agencies. Each of these participants has a different
perspective and is involved with different issues related to buildings energy use.

Consider automobiles as a counterexample. Improving automobile
fuel economy involves a few major manufacturers in Detroit (plus
several foreign car makers). Cars driven in one part of the country
are basically the same as those driven in other parts of the country,
except for slight variations due to climate, topography, and urban
congestion. On the other hand, buildings differ widely from region
to region because of such factors as variations in winter and summer
temperatures, solar insolation (sunshine), humidity, availability and
prices of different fuels, cost of land, and historical practices.
These regional factors suggest the difficulties of introducing conservation innovations into buildings.

A fourth barrier to full-scale adoption of energy efficiency measures concerns regulation. Almost every community in the United States has its own building code, originally established to ensure compliance with fire, health, and safety requirements. These codes theoretically offer an opportunity for increasing energy efficiency of new buildings by incorporating requirements related to the thermal performance of structures and the efficiency of HVAC systems. In reality, the multiplicity of building codes and the differences among the codes and how they are administered complicate and reduce this potential.

Moreover, when the owner of a building is also the sole occupant of that building, there is little ambiguity about who pays the costs and who received the benefits of adopting conservation measures. However, for a building that is tenant-occupied, this becomes an important issue. If the tenant pays for utilities, then the owner has little incentive to invest in energy-efficient systems: the costs would accrue to the owner, but the benefits would accrue to the tenant. On the other hand, the tenant has little incentive to invest in conservation. Although the tenant would enjoy lower fuel bills while he remained in that building, he would not recoup the investment when he left--another case in which the costs and benefits accrue to different parties. The fact that 35% of the nation's housing stock is occupied by renters suggests that this is an important--but unresolved--issue.

A final obstacle to more rapid energy conservation concerns uncertainty over the future. Decisions about the future are always complicated by uncertainty, but this is particularly true in the energy field as a result of recent changes in fuel prices and the problem of fuel availability. Decision-makers are likely to underinvest when faced with great uncertainty about future fuel prices and related government policies.

FUTURE ENERGY USE

In order to consider the aspects of our energy future, we must evaluate the energy and economic effects of changes in fuel prices and
the consequences of adopting various conservation programs in the
buildings sectors. The timespan considered is from the present to
the end of this century. Analyses are conducted using two detailed
engineering-economic models of energy use for each of the two building sectors; residential and commercial (10,11).

Four scenarios are examined: (1) a high case, in which real
fuel prices remain constant from 1976 to 2000; (2) a baseline
example, in which fuel prices rise over time; (3) a conservation
case that includes higher electricity and gas prices plus the regulatory, financial incentive, and information programs of both the
1978 National Energy Act and preceding federal legislation; and
(4) another conservation case that also includes the development
and introduction of new technologies (more efficient equipment,
appliances, and structures). Each scenario is evaluated in terms
of its effects on buildings energy use (by fuel, by end use, and in
aggregate) and on direct economics (fuel bills, capital costs for
equipment and structures) between now and the year 2000.

High Projection

Assuming that real fuel prices to the R/C sectors remain unchanged
from their 1976 values and assuming that the population grows according to the Bureau of the Census Series II projection (23), we
can develop projections of energy use to the year 2000. Per capita
income can be derived from a recent projection of GNP prepared for
DOE and the Series II population projection (24). Projections of
household formation, stocks of occupied housing, commercial floorspace, and new construction are from the models mentioned above
(10,11). Table 1-8 shows population, number of households, amount
of commercial floorspace, and per capita income used in all the
projections. Finally, in this case, we assume that there are no
government programs to encourage energy conservation (that is, we

Table 1-8. Inputs Used in Projections of Residential and Commercial Energy Use to 2000

	Population (millions)	Households (millions)	Commercial Floorspace (billion ft^2)	Per Capita Income (1975-$)
1970	204	63	24.3	5,420
1975	213	71	28.3	5,850
1976	215	73	29.2	6,050
1980	222	80	32.6	7,150
1985	233	88	38.4	7,970
1990	244	95	44.2	8,890
2000	260	106	57.0	10,570

ignore recent legislation; the likely effects of these programs are discussed later).

Outputs from the energy models, given these inputs, show energy use growing from 26 QBtu in 1977 to 29 QBtu in 1980, 38 QTbu in 1990, and 47 QBtu in 2000, with an average annual growth rate of 2.7%. Commercial fuel use grows much more rapidly (3.2% per year), primarily because growth in commercial floorspace is so much greater than is growth in households (2.9% versus 1.6% per year). In fact, on a per-unit basis (per square foot of commercial floorspace or per household for the residential sector), growth rates are similar for the two sectors. Growth in energy use is slower in the projection period than it has historically been (2.7% versus 4.0% per year). This reduction is due to the effects of projected demographic changes and higher fuel prices during the mid-1970s (see Fig. 1-3), of operational changes, and of technical efficiency improvements.

The contribution of different fuels to the total used changes during the projection period. Because of sharp increases in petroleum and gas prices during the mid-1970s, consumer preference for electricity, and a steady rise in incomes (which leads to greater ownership of air conditioners, refrigerators, freezers, lights, and small electrical appliances), the fraction of sector fuel use devoted

to electricity increases from 52% in 1976 to 61% in 2000. Shares
contributed by gas and oil decline from 29 and 16% to 27 and 11%,
respectively.

Baseline Projection

The baseline projection differs from the high projection only with
respect to future fuel prices. Rather than assume that real fuel
prices remain constant at their 1976 values, we assume that fuel
prices increase over time, as projected by the DOE and Brookhaven
National Laboratory. Figure 1-13 shows the projected fuel prices
of the residential sector (commercial sector prices follow much the
same trends). These curves show substantial increases in baseline
gas prices (3.3% per year), while oil (1.8% per year) and electricity (1.4% per year) prices increase more slowly.

In the baseline projection, buildings energy use grows from 26
QBtu in 1977 to 27 QBtu in 1980, 32 QBtu in 1990, and 38 QBtu in
2000, with an average annual growth rate of 1.7% (see Figs. 1-14 and
1-15). Again, commercial fuel use grows more rapidly than does residential fuel use: 2.2% versus 1.5% per year. Energy growth is

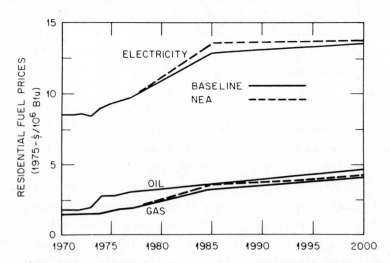

*Figure 1-13. Projected fuel prices with and without NEA, 1970-
2000.*

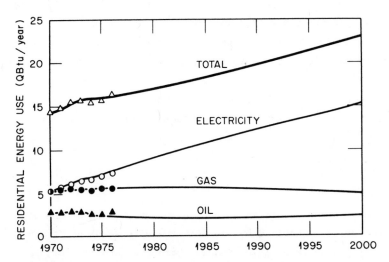

Figure 1-14. Baseline projection of residential energy use, 1970-2000.

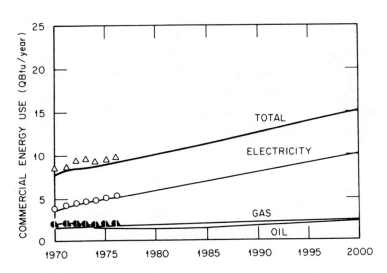

Figure 1-15. Baseline projection of commercial energy use, 1970-2000.

slower in the baseline projection than in the preceding case because
of voluntary responses to higher fuel prices. Energy use in buildings is reduced because of both operational changes and improvements
in the energy efficiencies of new equipment, new buildings, and
existing buildings.

Changes in the distribution of energy use by fuel are similar
to those in the high projection case. Electricity increases its
share of the total used, while the contributions of gas and oil both
decline over time.

Government Conservation Programs

The energy and economic effects on the buildings sectors of the conservation programs authorized by the 94th and 95th Sessions of Congress (25-27) during 1975, 1976, and 1978 are evaluated in this section. These programs involve:

1. Appliance efficiency standards (residential only) implemented
    in 1980.
2. Thermal performance standards adopted in 1978, for the construc-
    tion of new buildings, and more stringent standards for 1980.
3. Weatherization programs for existing buildings (retrofit):
    a. Federal grants to schools, hospitals, and low-income house-
        holds for weatherization of existing buildings.
    b. Federal tax credits for weatherization of residential and
        commercial buildings.
    c. Federal Energy Management Program to reduce energy use in
        federal buildings.
4. Higher prices for electricity and gas (see Fig. 1-13).

Appliance Efficiency Standards. The Department of Energy is responsible for developing and implementing a set of efficiency targets to
raise the average efficiency of new household appliances and equipment sold in 1980 at least 20% from the 1972 average (25). Thirteen
classes of appliances are considered, among which the more important
(from an energy use standpoint) are space heating equipment, water
heaters, refrigerators, freezers, and air conditioners. The NEA

made this program of voluntary targets into one of mandatory standards, with efficiency improvement targets ranging from about 10%
(oil furnaces) up to 60% (gas ranges/ovens) (28).

Performance Standards for New Construction. The DOE and the Department of Housing and Urban Development (HUD) are developing thermal
performance standards for the construction of all new buildings.
These standards must then be implemented by the individual states,
but only if Congress first approves them (26). President Carter
proposed to "advance by one year, from 1981 to 1980, the effective
date of the mandatory standards required for new residential and
commercial buildings" (29). These standards are likely to be somewhat more stringent than those developed by the American Society
of Heating, Refrigerating, and Air Conditioning Engineers (ASHRE
90-75) in 1975 (30,31).

Retrofit Programs. Recent legislation includes several programs to
encourage weatherization of existing residential and commercial
buildings (25-27). In the commercial sector, these programs consist
of a 10% investment tax credit, in effect from 1977 through 1982,
for efficiency improvements; a 3-year, $900 million grants program
for schools and hospitals; and the Federal Energy Management Program, developed to upgrade buildings operated by the federal government. The residential sector benefits from a 15% tax credit for
retrofit costs up to $2000, in effect from 1977 through 1984;
weatherization grants, which total $530 million for the period 1978-
1980 (32); to low-income families; several DOE and HUD demonstration programs; mandatory electric and gas utility programs (home
audit services); and a rural home weatherization program.

Higher Fuel Prices. Provisions in the NEA are likely to raise electricity prices (by requiring state regulatory commissions to consider several standards concerning rate design and other utility
practices) and raise natural gas prices (by requiring changes in
natural gas regulations) (27). The likely effect of these measures
on residential fuel prices is shown in Fig. 1-13.

Tables 1-9 and 1-10 summarize the energy and economic effects
of adopting these federal conservation programs; details on the effects of each individual program are contained in Refs. 33 and 34.
Energy savings relative to the baseline projection increase from
0.8 QBtu in 1980 to 2.0 QBtu in 1990 and 3.6 QBtu in 2000. Buildings
energy growth is cut from 1.7 to 1.3% per year. The comulative
energy savings between 1977 and 2000 is 57 QBtu, split almost equally
between the two sectors.

Table 1-10 shows the present worth (at a real interest rate of
8%) of changes from the baseline projection in energy-related expenditures. Fuel savings exceed the extra capital costs for improved
equipment and structures by $12 billion. Were it not for the higher
electricity and gas prices, net economic benefits would be much
greater. Relative to a "baseline" with the higher NEA fuel prices,
the NEA regulatory and financial incentive programs yield a net economic benefit of $35 billion to households (see footnote of Table
1-10).

*Table 1-9. Alternative Residential and Commercial Energy Projec-
tions: Energy Use*

	Energy Use (QBtu)				Average Annual
	1980	1985	1990	2000	Cumulative (1978-2000)	Growth Rate 1977-2000 (%)*
1. Constant Fuel Prices, No Government Programs						
Residential	18.2	20.8	23.5	28.1	524	2.3
Commercial	10.4	12.5	14.5	18.9	326	3.2
2. Baseline						
Residential	17.3	18.4	20.1	23.1	456	1.5
Commercial	9.5	10.5	11.8	14.8	270	2.2
3. Baseline plus NEA						
Residential	16.8	17.0	18.4	21.4	424	1.1
Commercial	9.2	9.7	10.7	12.9	245	1.7
4. Baseline, NEA, RD&D						
Residential	16.7	16.8	17.8	19.7	409	0.8
Commercial	9.2	9.6	10.4	12.2	241	1.3

*Table 1-10. Alternative Residential and Commercial
Energy Projections: Direct Economic Effects*

	Present Worth of Cumulative (1978-2000) Expenditures at 8% Real Interest Rate Relative to Baseline (billion 1975-$)			
	Fuels	Equipment	Structures	Net
1. Constant Fuel Prices, No Government Programs				
Residential	−140.0	−4.7	−2.2	−146.9
Commercial	−65.5	−2.4	−0.9	−68.8
2. Baseline				
Residential	0	0	0	0
Commercial	0	0	0	0
3. Baseline plus NEA*				
Residential	−22.6	8.3	14.3	0
Commercial	−16.0	4.2	0.2	−11.6
4. Baseline, NEA, RD&D*				
Residential	−41.2	10.4	11.3	−19.5
Commercial	−20.0	4.8	0.3	−14.9

*These numbers include both the costs due to higher fuel
prices and the savings due to the NEA conservation programs and new technologies. For example, the net economic benefit of the NEA programs relative to a baseline with the higher NEA fuel prices is $18 billion
for the residential sector and $17 billion for the com-
mercial sector.

New Technologies

Both private industry and the federal government (DOE) are conduct-

ing research, development, and demonstration (RD&D) programs to

bring to the market new systems for satisfying building energy-

related functions. These new systems are likely to be much more

energy-efficient than are the existing ones (35,36). For example,

gas-fired heat pumps are expected to satisfy annual space heating

requirements in typical buildings by using about half the natural

gas that conventional gas furnaces and boilers consume. Develop-

ment of improved control systems (especially for commercial build-

ings) that utilize sensors, actuators, controllers, and logic cir-

cuits are likely to reduce both equipment energy use and peak electric loads. These systems are able to sense the outdoor air temperature and enthalpy and use fresh air for air conditioning when appropriate. Such systems are also able to recirculate inside air
when a particular area in a building requires heating (e.g., the
north-facing perimeter) and a different area requires cooling (e.g.,
the core).

Figure 1-16 shows how the energy savings resulting from the NEA
programs discussed in the preceding section plus the new technologies
considered here grow over time. Savings resulting from new technologies alone increase from 0.1 QBtu in 1980 to 0.9 QBtu in 1990 and
2.4 QBtu in 2000. Energy consumption is cut from 1.3% per year
(with NEA programs) to 1.0% per year. The cumulative energy savings
between 1977 and 2000 is 19 QBtu (see Table 1-9).

For the residential and commercial sectors the net economic
benefit of these new technologies is $23 billion (see Table 1-10).
Fuel bills are reduced by (a net present worth of) $23 billion.
Equipment costs increase by $3 billion; however, costs for structures
decrease by $3 billion. This suggests that the technologies for
new structures (provided by RD&D programs) can reduce the cost of
meeting the NEA goals concerning construction of new buildings and
retrofit of existing ones.

RD&D benefits differ substantially from the benefits brought
about by the NEA, as shown for the residential sector in Fig. 1-17.
Energy savings resulting from higher gas and oil prices and from
the regulatory, financial incentive, and information programs of the
NEA increase rapidly through the early 1980s and then only slightly
from 1985 to 2000. RD&D benefits, on the other hand, grow slowly
at first, but then in the 1990s they grow much more rapidly than do
the NEA benefits. In the year 2000, the residential energy savings
resulting from RD&D (1.7 QBtu) is equal to the residential NEA savings; in 1985 the RD&D energy savings is only one-tenth of the NEA
savings.

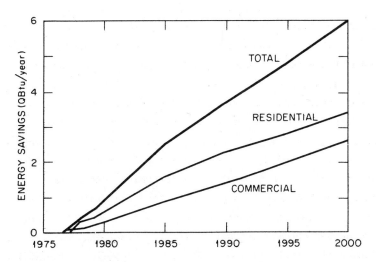

Figure 1-16. Projected savings in the buildings sector due to NEA and RD&D, 1975-2000.

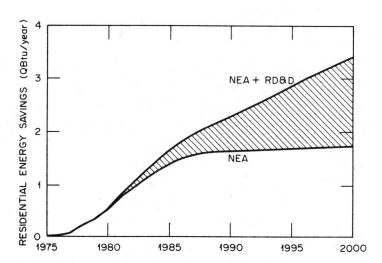

Figure 1-17. Projected residential energy savings due to NEA and RD&D, 1975-2000.

CONCLUSION

In this chapter we reviewed historical trends in buildings energy
use from 1950 to 1978, discussed several specific conservation
measures, noted the major obstacles to rapid and widespread adoption
of these measures, and presented four scenarios of future buildings
use to show the energy and economic effects of different assumptions
concerning government conservation programs, future fuel prices, and
future technologies. The major conclusion that can be drawn is that
the range of future energy use in residential and commercial buildings depends strongly on both private and government decisions concerning fuel prices, conservation programs, technical improvements,
and behavioral changes. To illustrate, for the year 2000, the range
in energy use across the limited set of futures considered here is
32 to 47 QBtu, a difference of about ±20% of the baseline value (37).

As shown previously, the range of conservation measures available today is impressive. A combination of minor operational changes
and cost-effective technical improvements could cut buildings energy
use by as much as one-half. Of course, adoption of these changes
takes time, as existing stock of equipment and structures wears out
and is replaced.

A comparison of the high projection with historical statistics
suggests that future buildings energy growth is likely to be much
lower than it has been in the past (4.2% per year between 1950 and
1973 and 2.7% per year between 1977 and 2000). The combined effects
of recent increases in fuel prices and of no further declines in fuel
prices (Fig. 1-3) plus slower growth in population, number of households, and amount of commercial floorspace (Table 1-8) will lead to
a substantially reduced growth in buildings energy use.

The effects of projected fuel price increases (Fig. 1-12) reduce buildings energy growth from 2.7% per year in the high case to
1.7% per year in the baseline case. Thus, the free market (voluntary) response to rising fuel prices is expected to lead to a slower
energy growth than has been in effect historically. The baseline
estimate of buildings energy use for the year 2000 (38 QBtu) is about

half of what would occur had preembargo energy growth rates continued through the end of the century.

Starting from this baseline of slower energy growth, it is perhaps surprising to realize that government conservation programs and development of new buildings technologies further reduce energy consumption: 6 QBtu in 2000, 16% of the baseline figure. The cumulative energy savings between 1977 and 2000 is 76 QBtu, equal to 3 years of present-day buildings energy use. In addition, conservation programs and new technologies offer large economic benefits to occupants of residential and commercial buildings. The present worth of fuel bill reductions is likely to exceed the present worth of extra capital costs by $34 billion; the ratio of fuel bill reductions to extra capital costs (benefit/cost ratio) for these programs is 2.3. The real interest rate would have to be very much higher than the 8% used here before costs would exceed benefits.

Although our results show the technical and economic feasibility of energy conservation programs, it is not clear that the estimated energy and dollar savings will be easily realized. These programs require strong public support; dynamic, cost-effective, and timely regulations to increase the efficiencies of new equipment and structure; continued government and private RD&D to develop and produce improved buildings technologies; and active cooperation among manufacturers and from organizations involved in the design, construction, and financing of structures (architects, builders, contractors, suppliers, banks). All these "requirements" suggest that it may be difficult to achieve estimated energy and economic benefits.

Nevertheless, it is clear that regulatory, incentive, informational, and RD&D programs for energy conservation can substantially reduce buildings energy use between now and the end of the century. Such programs also save money, reduce the environmental effects of energy production and use, and provide more time for the development of new energy supplies.

NOTES AND REFERENCES

1.   1 QBtu = 1 Quad = 10^{15} Btu. Electricity use figures are in
     terms of primary energy (11,500 Btu/kWh) to include losses in
     generation, transmission, and distribution.

2.   All prices are given in terms of 1975 dollars. We use "con-
     stant" dollars to correct for the effects of inflation.

3.   Bureau of the Census, *Statistical Abstract of the United States:
     1978*, U.S. Department of Commerce, September 1978.

4.   Bureau of the Census, *Annual Housing Survey: 1976*, Part A,
     Series H-150-76, U.S. Department of Commerce and U.S. Depart-
     ment of Housing and Urban Development, 1978; also earlier
     annual issues.

5.   J. R. Jackson and W. S. Johnson, *Commercial Energy Use: A
     Disaggregation by Fuel, Building Type and End Use*, ORNL/CON-14,
     Oak Ridge National Laboratory, Oak Ridge, Tenn., February 1978.

6.   Council of Economic Advisers, *Economic Report of the President*,
     January 1979.

7.   Bureau of Mines, Annual U.S. Energy Up in 1976, press release,
     U.S. Department of Interior, Mar. 14, 1977. Also, C. Readling,
     personal communication, U.S. Department of Energy, June 26,
     1978. Also, E. Hirst and J. Jackson, Historical patterns of
     residential and commercial energy uses, *Energy*, vol. 2, no. 2,
     June 1977.

8.   Bureau of Labor Statistics, *Retail Prices and Indexes of Fuels
     and Utilities*, monthly issues, U.S. Department of Labor, 1950-
     1979.

9.   S. Cohn, *Fuel Choice and Aggregate Energy Demand in the Com-
     mercial Sector*, ORNL/CON-27, Oak Ridge National Laboratory,
     Oak Ridge, Tenn., December 1978.

10. J. R. Jackson, et al., *The Commercial Demand for Energy: A
     Disaggregated Approach*, ORNL/CON-15, Oak Ridge National Labora-
     tory, Oak Ridge, Tenn., April 1978.

11. E. Hirst and J. Carney, *The ORNL Engineering-Economic Model of
     Residential Energy Use*, ORNL/CON-24, Oak Ridge National Labora-
     tory, Oak Ridge, Tenn., July 1978.

12. D. Quentzel, Night-time thermostat set back: Fuel savings in
     residential heating, ASHRAE J., March 1976. Also, D. A.
     Pilati, *The Energy Conservation Potential of Winter Thermostat
     Reduction and Night Setback*, ORNL-NSF-EP-80, Oak Ridge National
     Laboratory, Oak Ridge, Tenn., February 1975.

13. J. C. Moyers, *The Room Air Conditioner as an Energy Consumer*,
     ORNL-NSF-EP-59, Oak Ridge National Laboratory, Oak Ridge,
     Tenn., October 1973.

14. D. A. Pilati, *Room Air Conditioner Lifetime Cost Considerations: Annual Operating Hours and Efficiencies,* ORNL-NSF-EP-85, Oak Ridge National Laboratory, Oak Ridge, Tenn., October 1975.

15. W. S. Johnson and F. E. Pierce, *Energy and Cost Analysis of Commercial Building Shell Characteristics and Operating Schedules,* ORNL-CON-39, Oak Ridge National Laboratory, Oak Ridge, Tenn., April 1980.

16. J. P. Acton, M. H. Graubard, and D. J. Weinschrott, *Electricity Conservation Measures in the Commercial Sector: The Los Angeles Experience,* R-1592-FEA, Rand Corporation, Santa Monica, Cal., September 1974.

17. P. F. Hutchins, Jr. and E. Hirst, *Engineering-Economic Analysis of Single-Family Dwelling Thermal Performance,* ORNL/CON-35, Oak Ridge National Laboratory, Oak Ridge, Tenn., November 1978.

18. D. L. O'Neal, *Energy and Cost Analysis of Residential Space Heating Systems,* ORNL/CON-25, Oak Ridge National Laboratory, Oak Ridge, Tenn., July 1978.

19. R. A. Hoskins, E. Hirst, and W. S. Johnson, Residential refrigerators: Energy conservation and economics, *Energy,* vol. 3, no. 1, February 1978.

20. R. E. Lyman, Jr. and W. S. Johnson, *Energy and Cost Analysis of Commercial HVAC Equipment: Office and Hospitals,* University of Tennessee, Knoxville, Tenn., March 1979.

21. R. L. Dunning, E. R. Amthor, and E. J. Doyle, *Research and Development of a Heat Pump Water Heater,* vol. 1 (ORNL/Sub-7321/1) and vol. 2 (ORNL/Sub-732/2), Energy Utilization Systems, Inc., Pittsburgh, Penn., August 1978. Also, D. O'Neal, E. Hirst, and J. Carney, *Regional Analysis of Residential Water Heating Options: Energy Use and Economic,* ORNL/CON-31, Oak Ridge National Laboratory, Oak Ridge, Tenn., October 1978.

22. E. Hirst, Effects of the national energy act on energy use and economics in residential and commercial buildings, *Ener. Sys. Policy,* vol. 3, no. 2, 1979.

23. Bureau of the Census, Projections of the population of the United States: 1977-2050, *Current Population Reports,* Series P-25, No. 704, U.S. Department of Commerce, July 1977.

24. S. Carhart, personal communication, Brookhaven National Laboratory, Jan. 17, 1978 and Mar. 15, 1978. Also R. Sastry, personal communication, U.S. Department of Energy, Jan. 27, 1978.

25. 94th Congress, Energy Policy and Conservation Act, P.L. 94-163, Dec. 22, 1975.

26. 94th Congress, Energy Conservation and Production Act, P.L. 94-385, Aug. 14, 1976.

27. 95th Congress, National Energy Act; Nov. 9, 1978: Public Util-
    ities Regulatory Policies Act, P.L. 95-617; Energy Tax Act of
    1978, P.L. 95-618; National Energy Conservation Policy Act of
    1978, P.L. 95-619; National Gas Policy Act of 1978, P.L. 95-621.

28. Federal Energy Administration, Energy conservation program for
    appliances, *Fed. Reg.*, vol. 43, no. 70, Apr. 11, 1978. Also,
    Fed. Reg., vol. 43, no. 198, Oct. 12, 1978.

29. The White House, *The President's Energy Program*, Apr. 20, 1977.
    Also, Executive Office of the President, *The National Energy
    Plan*, Apr. 29, 1977.

30. American Society of Heating, Refrigerating, and Air Condition-
    ing Engineers, Energy Conservation in New Building Design,
    ASHRAE 90-75, 1975. Also, Arthur D. Little, Inc., *An Impact
    Assessment of ASHRAE Standard 90-75, Energy Conservation in New
    Building Design,* Cambridge, Mass., December 1975.

31. U.S. Department of Energy, Energy performance standards for new
    buildings, *Fed. Reg.*, vol. 43, no. 255, Nov. 21, 1978. Also,
    AIA Research Corporation, Phase 2 Report: Development of Energy
    Performance Standards for New Buildings, draft, November 1978.

32. U.S. General Accounting Office, Complications in Implementing
    Home Weatherization Programs for the Poor, HRD-78-149, August
    1978.

33. E. Hirst and J. Carney, Effects of federal residential energy
    conservation programs, *Science,* vol. 199, no. 4331, Feb. 24, 1978.

34. J. R. Jackson, *An Economic-Engineering Analysis of Federal E
    Energy Conservation Programs in the Commercial Sector,* ORNL/CON-
    30, Oak Ridge National Laboratory, Oak Ridge, Tenn., January 1979.

35. Division of Buildings and Community Systems, Buildings and Com-
    munity Systems Program Approval Document: FY 1978, Energy Re-
    search and Development Administration, August 1977. Also,
    Division of Buildings and Community Systems, Consumer Products
    and Technology Branch Program Plan, ERDA 77-81, Energy Re-
    search and Development Administration, September 1977.

36. J. H. Gibbons, Long-term research opportunities, in *Energy Con-
    servation and Public Policy,* Prentice-Hall, 1979, pp. 210-228.

37. The range of feasible solutions for the year 2000 is actually
    much greater than those covered by the four scenarios presented
    here. For example, it is possible that new building end uses
    (e.g., sidewalk de-icing, inside air filtration, swimming pool
    heating) will appear between now and the end of the century and
    will substantially increase energy use beyond the level of the
    high projection. On the other hand, our projections do not in-
    clude the possibility of widespread use of solar systems for
    space and water heating, nor do they include use of community
    energy systems that provide both heat and electricity. Per-
    haps there are other cost-effective energy conservation measures
    that are also not included in our projections.

2
Lighting

*Robert T. Dorsey**

Lighting Technology Development
Lamp Marketing and Sales Operation
General Electric Lighting Business Group
Nela Park, Cleveland, Ohio

We use lights and lighting systems to help us do our jobs in the most
productive way possible. But because the lighting systems of many
existing buildings were designed within the restrictions of initial
cost economies, without knowledge about final space use and subdivi-
sion, and without benefit of relatively recent developments, there
exists significant potential for lighting system modifications.
These modifications can reduce substantially the energy consumed by
the lighting system, while still providing the quality of illumination required to perform all necessary tasks and functions.

However, before undertaking any change it should be recognized
that a lighting system is just that--a system. Its many elements
are interrelated with one another, just as the lighting system itself is interrelated with other systems in the building. Although
energy can be conserved by properly removing lamps and luminaires,
such action should be taken only after the entire system has been
analyzed and all options evaluated. Conservation of energy is important, but it must be achieved in a manner consistent with pro-

*Present affiliation: Professional Engineer, 3726 Tolland Road,
Shaker Heights, Ohio

ductivity and visual comfort, with aesthetics, and within federal,
state, and local codes and ordinances. It is also important to
recognize that major alterations to a lighting system can have a
significant impact on heating and cooling systems, many of which
were designed to make use of or offset the amount of heat given off
by the lighting system.

Taking all of these factors into consideration, this chapter
discusses in some detail the effective management of lighting in
buildings. It also discusses the impact of lighting on productivity
and provides recommendations for improved lighting management.

PUTTING LIGHTING IN PERSPECTIVE

The total power generated in 1978 was about 2.3×10^{12} kWh. Of this,
16%, or about 368×10^9 kWh, was for lighting, which translates to
about 657 million barrels of oil-equivalent per year, or to about
1.8 million barrels per day (BPD). Figure 2-1 shows the different
energy applications as percentages of the total energy demand. Three
important applications have been factored out of the three consumer
areas (industrial, residential, and commercial-institutional), so
that they might be discussed and seen in perspective. These are
lighting (5%), heating (24%), and cooling (3%).

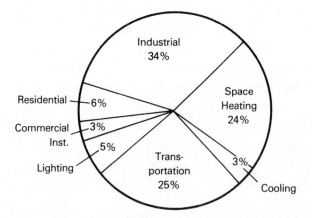

Figure 2-1. Energy applications. (From U.S. Department of Energy.)

Table 2-1 presents the amounts of energy used for lighting in its various applications. On a short-term basis, conserving 25% of this energy would amount to a saving of 450,000 BPD. Sometime in the mid-1980s, this savings figure might be as much as 40%, which would translate to 720,000 BPD, and if the figure reached 50% in the year 2000, it would mean a 900,000 BPD savings.

Detailed data on lighting energy usage by watts per square foot in various building types are given in Table 2-2. According to an American Institute of Architects (AIA) Research Corporation report, average lighting consumes about 29% of the energy requirement of a building. This amounts to an average of 2.8 W/ft^2, 3376 hr of operation per year, and 112,000 total Btu/ft^2 consumption.

LIGHTING AND PRODUCTIVITY

The relationship between light, vision, visibility, visual task, the worker, and his or her productivity is an important one. Therefore, a cost/benefit analysis of the entire visual system, not just of the artificial lighting portion of it, can determine the method for

Table 2-1. Lighting Applications

	% of Lighting KWH	Total Energy Consumed in U.S.
Residential	20%	1%
Industrial	20%	1%
Store	20%	1%
Street and Highway	3%	.2%
Flood, Area, Signs, Sports, Etc.	8%	.4%
Offices	8%	.4%
Schools	7%	.3%
All Other Indoor	14%	.7%
	100%	5.0%

Table 2-2. Lighting (W/ft^2)

Building Type	Total No. of Bldgs.	Total Sq. Ft. (1000 SF)	Percent of Total Sq. Ft. With Average Watts/Sq. Ft. in Each Category								Summary Statistics			
			<1	1-<2	2-<3	3-<4	4-<5	5+	Not Spec.	Tot %	Range From	To	Avg.	St. Dev.
Office	237	9847.7	0.0	13.3	34.8	35.1	16.6	0.2	0.0	100.0	0.7	6.6	2.8	0.8
Education - Elementary	157	5695.5	0.0	11.2	58.2	28.2	2.5	0.0	0.0	100.0	1.2	4.6	2.6	0.6
Education - Secondary	172	1334.9	0.4	11.8	56.5	24.5	6.3	0.5	0.0	100.0	0.5	5.0	2.6	0.7
Education - College/Univ.	57	3895.6	0.0	10.2	57.8	26.7	3.1	2.2	0.0	100.0	1.3	7.5	2.8	1.0
Hospital	40	6754.8	0.0	11.7	47.1	35.8	5.4	0.0	0.0	100.0	1.5	4.0	2.6	0.6
Clinic	113	3508.5	0.0	5.4	45.7	40.9	8.0	0.0	0.0	100.0	1.4	4.5	2.8	0.7
Assembly	167	2734.9	1.8	36.7	41.2	15.6	4.1	0.7	0.0	100.0	0.3	9.2	2.2	1.0
Restaurant	196	1003.8	1.8	7.6	57.4	24.3	6.1	2.2	0.6	100.0	0.9	5.4	2.7	0.9
Mercantile	176	9276.5	0.1	22.9	40.7	17.8	11.6	5.8	1.2	100.0	0.7	8.4	2.7	1.1
Warehouse	81	2401.4	12.3	44.4	22.4	20.9	0.0	0.0	0.0	100.0	0.2	3.9	1.8	0.8
Residential Non-Housekeeping	162	8736.0	0.2	38.5	47.3	12.8	0.9	0.3	0.0	100.0	0.6	6.4	2.1	0.8
TOTALS	1557	67198.6	0.6	18.7	46.9	25.5	6.9	1.1	0.3	100.0				

achieving optimum performance in a task environment which is constrained by use of energy for lighting. But in order to develop the
necessary data for such a cost/benefit analysis, we must have some
knowledge about the way in which human performance is affected by
the quantity and quality of illumination and about alternate means
of improving performance (e.g., changing the task itself).

As it so happens, the financial impact of raising or lowering
productivity levels can be highly significant, as was demonstrated
in several controlled tests. In one, a study of the performance of
keypunch operators at the General Electric Company, energy consumption for lighting was reduced by lowering the level of illumination
from 150 fc (footcandles) to 50 fc (see Table 2-3). As a result, the
time it took to get the same amount of work done increased by 13.6%.
At 150 fc, the cost of labor on a square-foot basis was $91 per year;
at 50 fc, it rose to $102.59 per year. Thus, while lowering the il-
lumination level saved $0.70/ft^2/year, the reduced productivity re-
sulted in an overall net *loss* of $11.54/ft^2/year (1). In addition,
when the illumination level was reduced from 100 to 50 fc, the productivity rate fell by 28%. Productivity jumped when the 100-fc
lighting level reduction was restored, then receded with time to the
original rate.

In another experiment, designed by the Federal Energy Administration and conducted at the Vision Research Laboratories at Ohio
State University, handwritten numbers, some of which were wrong,

Table 2-3. *Cost Analysis Based on Time Required to Perform Work*
When Lighting is Reduced from 150 to 50 fc

FC	Relative Time for Same Work	Labor Cost for Same Work ($/Sq. Ft./Yr.)	Total Cost of Light ($/Sq. Ft./Yr.)	Total Net Cost ($/Sq. Ft./Yr.)	Net Loss ($/Sq. Ft./Yr.)
150	1.0000	$ 90.00	$1.05	$ 91.05	
50	1.136	$102.24	$0.35	$102.59	
		$ 12.24	$0.70		$11.54

were compared to a correct checklist (2). The lighting levels were
varied randomly; the variables that affect productivity were carefully controlled. The results indicate that it takes less time to
do the same amount of work under better vision conditions.

Table 2-4 provides the data on productivity and costs at three
lighting levels. As can be seen from the table; the increased productivity of the greater lighting levels resulted in reduced total
output cost. In this analysis, which is based on the 1976-1977
Administrative Management Society Office Salaries Survey, the average office space per worker is assumed to be 100 ft^2, and the cost
of electricity is 3½¢/kWh.

Poor lighting can keep one from performing a task in the most
effective manner. Consequently, when evaluating the lighting requirement by the task, one should consider four fundamental factors.
The first of these is the *time* it takes to see: adequate light helps
a worker complete the task faster, without errors. The second factor is *size* of the detail. The smaller something is, the harder it
is to see, as is evident in reading small print. The third factor
is the *contrast* between details and their background. Materials with
low or poor contrast, such as multiple carbon copies, are harder to
see than those with high contrast such as original typed material.
The last factor is *brightness*, or the amount of light reflected from
an object. Because the first three factors are often set by the
type of work that has to be done, brightness is usually the most

Table 2-4. *Productivity and Cost Analysis on Government Check Reading Study for Three Lighting Levels*

FC	Relative Time For Same Work	Labor Cost for Same Work ($/Sq. Ft./Yr.)	Total Cost of Light ($/Sq. Ft./Yr.)	Total Net Cost ($/Sq. Ft./Yr.)
50	1,0000	$ 91.00	0.40	$ 91.40
100	0.9485	$ 86.31	0.80	$ 87.11
150	0.9103	$ 82.84	1.20	$ 84.04

controllable factor. By using more light for the performance of a task, brightness or reflected light can be improved, which would in turn improve task visibility.

Lighting and Age

Different age groups need different amounts of light. As we get older, our need for light may change as a result of physiological changes in our eyes. Some older people need more light than younger people, and there may be an enormous difference between them in visual capabilities. With better visibility, people expend less effort in task performance; conversely, if visibility is poor, people feel that they should expend more effort. The next five charts summarize the relationship between task difficulty, illuminance, relative performance, the age of the worker, and the demands of the task shown upon the worker.

Figures 2-2 and 2-3 depict the effect of illuminance and task difficulty on the visual performance of young workers, age. 25. At relatively low task demand levels, easy tasks reach the point of diminishing returns at fairly low light levels, whereas moderately difficult tasks show significant improvement in performance at higher light levels. At higher task difficulty levels and with high task demand, easy tasks show gains in performance only with higher levels of illuminance. Thus, the figures show that moderate and difficult tasks cannot be raised to the same high levels of performance to which easy tasks can be raised, even with an increase in the level of illuminance.

Figures 2-4 and 2-5 depict the effect of illuminance and task difficulty on the visual performance of older workers. At relatively low task demand levels, low light levels result in substantial performance losses. At a high task demand level, lack of illumination can result in problems which are as yet little recognized. Simply stated, older workers doing demanding tasks that are difficult to see cannot obtain high levels of visual performance. In some cases, these workers can compensate for poor visual performance

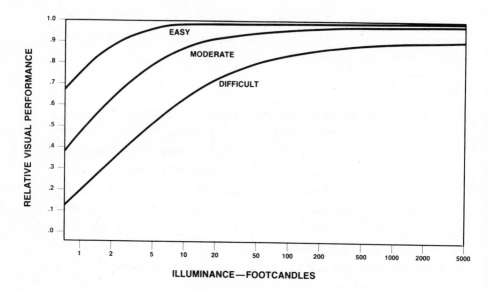

Figure 2-2. *Effect of illuminance and task difficulty on visual performance; age 25, task demand level 45.0. (Data from International Commission on Illumination 19/2.)*

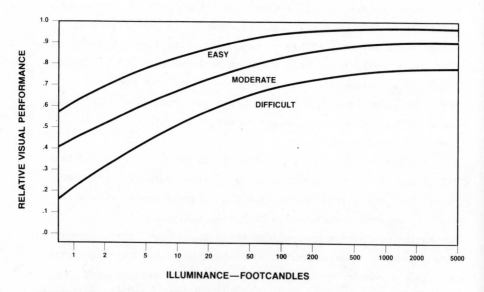

Figure 2-3. *Effect of illuminance and task difficulty on visual performance; age 25, task demand level 75.0. (Data from International Commission on Illumination 19/2.)*

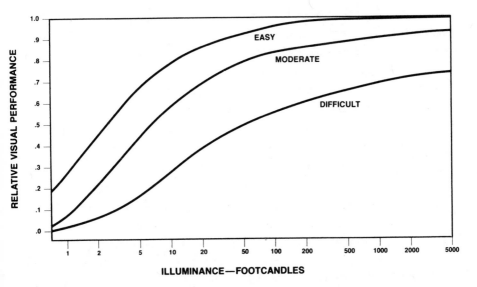

Figure 2-4. Effect of illuminance and task difficulty on visual performance; age 65, task demand level 45.0. (Data from International Commission on Illumination 19/2.)

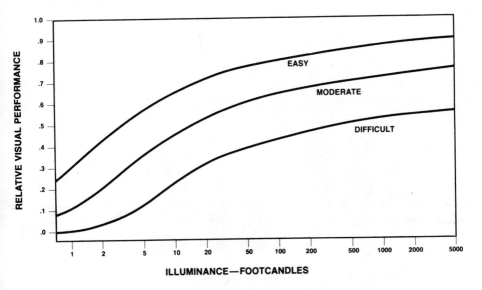

Figure 2-5. Effect of illuminance and task difficulty on visual performance; age 65, task demand level 75.0. (Data from International Commission on Illumination 19/2.)

by relying on their greater experience and on a sometimes stronger motivation for getting the job done. Unfortunately, these latter two factors are not always present.

Figure 2-6, which provides an indication of clerical workers' reactions to varying lighting levels, shows that older workers perceive their working conditions less favorably under lower levels of illumination than do younger workers.

Naturally, reducing the illumination level does not always affect productivity, since many systems now provide more light than is needed. In these cases, lighting levels should be reduced and the amount by which they are reduced require study in order to ensure that the modifications will not affect the ease of seeing along with other benefits provided by effective illumination.

Contrast Rendition

Contrast rendition refers to the relationship between a task and the task surface. Contrast rendition is affected by veiling reflections, which occur when illumination from an external source strikes the task surface in such a way that light bounces up into the viewer's eyes (see Fig. 2-7). Although the viewer may be completely unaware

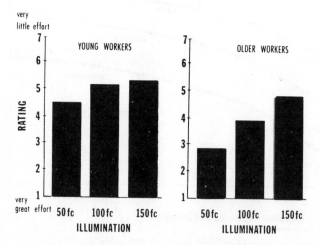

Figure 2-6. Clerical workers reactions to lighting levels.

Figure 2-7. Veiling reflection.

of this reflected light, it has the effect of reducing contrast
rendition, as if a thin, gauzy veil had been placed between the
viewer's eyes and the task surface.

Increasing the illumination level tends to improve contrast
rendition. But contrast rendition can also be improved without
raising the illumination level, by upgrading the quality of the illumination through more effective lamps, lenses, and luminaires.
Thus, improving contrast rendition can have the same effect on the
worker's vision as increasing the illumination level, without using
more energy to do so.

Because of the intimate relationship between quality and quantity of light, a new light-measuring concept has been developed.
Called equivalent spherical illumination, or ESI, it considers both
quality (in terms of veiling reflections) and quantity in the same
measure: ESI footcandles. ESI derives its name from the laboratory
test procedures developed by the Illuminating Engineering Society
to establish ESI lighting levels. In essence, a sphere source--a
uniformly bright hemisphere--is placed over a task situation to provide a standard comparison with the veiling reflections found in a
typical office space. In some cases, a task that may require 100 fc
from an actual lighting fixture may need only 20 fc of sphere illumination, since the sphere eliminates veiling reflections and the
office fixture does not.

By being more aware of the trigonometric relationships between
the light source, task surfaces, and the eye involved, lighting system designers can develop systems that minimize veiling reflections
and help keep the amount of artificial illumination actually needed
to a minimum.

Visual Comfort Probability

Visual comfort probability (VCP) is a measure that indicates how
much direct glare a luminaire (fixture) is likely to produce. Direct
glare comes from the light source within the normal field of view.
A VCP rating of 70--generally considered to be the lowest accept-
able--indicates that 70% of the people in a given space will not be
affected by discomfort glare.

Control of glare is important for workers in factories, and
particularly important for workers near windows. Choosing luminaires that have upward light components and painting the ceiling
white to reflect the light can greatly reduce the contrast between
luminaire and ceiling. And because the upward light component allows fixtures to stay cleaner, lighting efficiency is improved.

ENERGY CONSERVATION PRACTICES

Reducing energy waste while retaining the benefits that good illumination provides can be achieved in a number of ways, some of which may
even improve the quality of illumination. First of all, a comprehensive survey of lighting in a particular space should be conducted
in order to identify where the lights are and what kinds of lamps,
lenses, and luminaires are being used. The nature of the tasks to
be performed in the illuminated areas and the duration of these
tasks should also be considered, and existing lighting levels should
be compared with those that are recommended for the tasks involved.
A few general recommendations are given in Table 2-5.

Table 2-5. *Illuminance Categories and Values for Generic Types of Activities in Interiors*

Type of Activity	Illuminance Category	Ranges of Illuminances		Reference Work-Plane
		Lux	Footcandles	
Public spaces with dark surroundings	A	20-30-50	2-3-5	General lighting throughout spaces
Simple orientation for short temporary visits	B	50-75-100	5-7.5-10	
Working spaces where visual tasks are only occasionally performed	C	100-150-200	10-15-20	
Performance of visual tasks of high contrast or large size	D	200-300-500	20-30-50	Illuminance on task
Performance of visual tasks of medium contrast or small size	E	500-750-1000	50-75-100	
Performance of visual tasks of low contrast or very small size	F	1000-1500-2000	100-150-200	
Performance of visual tasks of low contrast and very small size over a prolonged period	G	2000-3000-5000	200-300-500	
Performance of very prolonged and exacting visual tasks	H	5000-7500-10000	500-750-1000	Illuminance on task, obtained by a combination of general and local (supplementary lighting)
Performance of very special visual tasks of extremely low contrast and small size	I	10000-15000-20000	1000-1500-2000	

The above tabulation is a consolidated listing of current illuminance recommendations. This listing is intended to guide the lighting designer in selecting an appropriate illuminance for design and evaluation of lighting systems. For illuminance category letters for specific areas and activities and guidance for the selection of an illuminance value within a specific range, see the *IES Lighting Handbook, 1981 Application Volume*, p. 2-4.

Source: *IES Lighting Handbook, 1981 Application Volume.*

Turn Off Lights

The energy consumed by lighting is a function of the number of watts used and the length of time a luminaire is operated. In other words, an electric bill depends on the amount of power used. One can save money with the obvious first step: Turn off the light when it is not needed.

Figure 2-8 illustrates the energy savings achieved by turning off lights during unoccupied periods for an office lighting system of 1000 fluorescent luminaires. The full system uses 92 kW. The upper part of the graph shows typical usage when lights are permitted to burn continuously from about 8:30 a.m. to 7:30 p.m. The shaded part of the graph shows more efficient use of the same system: leaving lights on only when used. The energy savings amounts to 267.5

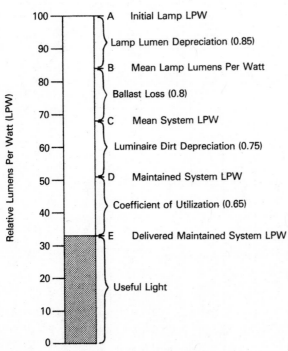

Figure 2-8. Energy savings achieved by turning off lights during unoccupied hours. (From Ref. 2.)

kWh/day, or 5875 kWh/month. At $0.04/kWh, this corresponds to a savings of $235 per month, or $2800 per year (3).

Of course, to make this possible there must be a switch available to the occupant's control. Obviously, if there is only one switch for a whole floor, individual employees cannot do much to help save money or energy. In designing a new building, switching should be planned so that only a few essential lights, rather than the number needed for normal building operation, can be left on when the room is unoccupied.

Adjust Illumination Levels

When a lighting survey shows that some people are getting either more or less illumination than they need for the tasks they perform, illumination levels must be adjusted. If too much light is being provided, one can change to a lamp that puts out less light, keeping in mind that the lamp should still be a high-efficiency lamp. If higher illumination levels are required, one can change to lamps with more output.

The building owner today has a wide spectrum of light sources from which to choose so as to be able to reduce lamp wattage and still maintain adequate illumination on the task. For example, the 150-W reflector lamps used in many downlights can be replaced by ellipsoidal reflector lamps which, in the 75-W size, can produce as much light as the 150-W size. Figure 2-9 shows a typical installation in which ellipsoidal reflector lamps achieve equal lighting levels with one-half the usual energy consumption while providing quality lighting for comfort and safety. In this 43-story building, fourteen 150-W reflector floodlamps were installed in recessed baffled downlights in each of its 43 elevator lobbies. By switching to the ellipsoidal reflector lamps, ER-30, a net savings of over $5000 annually was achieved, and the same level of illumination was maintained as with normal 150-W reflector lamps.

The reason greater illumination is provided is illustrated in Fig. 2-10. Sixty percent of the light from an R-20-type lamp is trapped in the fixture, whereas most of the light from an ER-30

Figure 2-9. Office using ellipsoidal reflector lamps. (Courtesy of General Electric, Cleveland, Ohio.)

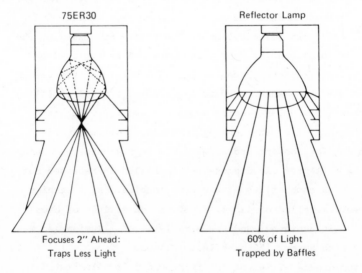

Figure 2-10. Ellipsoidal reflector lamp. (Courtesy of General Electric.)

ellipsoidal lamp is emitted in a criss-cross pattern that allows it
to escape the recess in which it is installed.

When two-lamp fluorescent fixtures are mounted in a row, one
should consider removing lamps in alternate fixtures of the row.
By doing this, instead of removing an entire row, better quality
lighting results. But, in order to maintain the recommended lighting level after removing the lamps, it may be necessary to use increased output lamps in place of some of those remaining. This
measure should still result in a net savings of energy. In many
cases it will be possible to relamp the remaining fixtures with more
efficient lamps to provide more lumens per watt. For example, 40-W
fluorescent tubes can now be replaced with 35-W tubes that produce
as much or more light, yet save 14 to 20% on energy consumption.
Table 2-6 provides an example of this substitution, which has resulted in annual savings of $2240 and a pay-back period of 2.4 years.

LIGHT SOURCES

Today there are so many different light sources available that being
able to choose the right energy-efficient light source for a given
task should present no problem. Not only are there more varieties,
there are new groups of lamps with electrical and light output characteristics far superior to the familiar incandescent and fluorescent lamps.

For use in indoor commercial and industrial applications, lamps
can be divided into several categories: incandescent, fluorescent,
and high-intensity discharge (mercury vapor, metal halide, high-
pressure sodium, and low-pressure sodium). Their comparative efficiencies are shown in Fig. 2-11. (HES designates low-pressure
sodium as "miscellaneous sources.")

High-intensity discharge, or HID, is the term commonly used to
designate four distinct types of lamps that actually have very little
in common. Of these four types--the mercury vapor, metal halide,
high-pressure sodium, and low-pressure sodium lamps--the high-
pressure sodium (HPS) lamp is the most efficient among those that
are normally used indoors. Because the sodium in the lamp is pres-

Table 2-6.   *Conversion to Maxi-Miser for 500 Troffers*

Capital Costs:		
2000 Watt-Miser II Lamps*		$ 3,500
1000 Maxi-Miser II Ballasts*		$11,000
Labor To Replace Ballasts & Lamps		$ 6,000
Total Investment		**$20,500**

Annual Operating Costs (Short Term):

	Standard System	New Maxi-Miser System
Energy	$14,080	$11,840
Ballast Replacement*	$ 3,700	(No Lamp or Ballast Replacements in First 2-3 Years After Conversion)
Lamp Replacement** (Lamp - $1.25, Labor $5.00)	$ 2,500	
Total	$20,280	$11,840
Saving		$8,440

Pay Back:

$$\frac{\$20,500 \text{ Investment}}{\$ 8,440 \text{ Annual Saving}} = 2.4 \text{ Years}$$

Continued Annual Energy Saving:
$14,080 - $11,840 = $2,240
Ballast Performance:
At Least Twice the Life of Standard Ballasts
Lighting:
4% More Light

*Max-Miser and Watt-Miser are trademarks of General Electric.

**10% burnouts per year.

***20% burnouts per year.

Source: Data from General Electric.

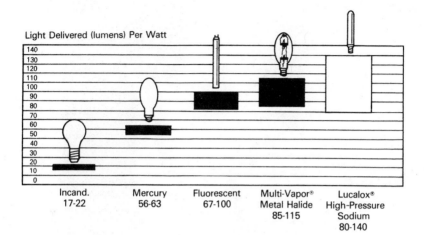

Figure 2-11. Light efficiencies. (Data from General Electric.)

surized, the light produced when electricity passes through the sodium vapor is not the characteristic bright yellow associated with sodium; rather, the light is "golden white" in color.

Figure 2-12 illustrates an example in which three hundred-twenty 1000-W vapor fixtures were replaced by one hundred forty-eight 1000-W HPS fixtures. The fixtures were mounted on a 30 × 30 ft^2 grid, 40 ft above the hangar floor. Although 100 fc of illumination was maintained, the changeover resulted in a 50% reduction in energy consumption and in an annual savings of more than $33,900.

The efficiency of a lamp is rated in lumens (light output) per watt (energy input). The higher the lumens per watt (LPW), the more light produced for the energy used. Obviously, using the most efficient kind of lamp or the most efficient version of a given kind of lamp minimizes energy consumption and operating cost. But it is not all that simple.

Of principal importance in lamp efficiency is the difference between the lamp's rate of initial light output and the amount of light that is actually effectively delivered to the area to be lighted. Figure 2-13 describes the difference between rated output

Figure 2-12. High-pressure sodium lighting. (Courtesy of General Electric.)

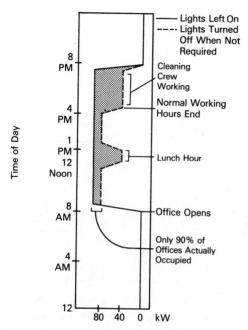

Figure 2-13.   Lamp depreciation characteristics.

and delivered output.   In relative terms, the initial lamp LPW is 100% at point A on the chart.   Over the life of the lamp, efficiency depreciates.   Since this rate of depreciation varies among the different lamps, an 85% depreciation was arbitrarily chosen for this chart, which lowers the initial efficiency (point A) to the mean efficiency (point B).

With the exception of incandescent lighting, light sources require a ballast to provide the proper voltage, wave shape, and current necessary to operate the lamp.   There is, inevitably, a degree of efficiency loss associated with the ballast device; which we have arbitrarily rated as 0.8.   This lowers the mean lamp efficiency (point B) to the mean system efficiency (point C).   Over a period of time, dirt collects on the luminaires and the lamp.   Because the amount varies widely, depending on location (e.g., on whether the

lighting system is in a foundry or an air conditioned office building), a depreciation value of 0.75 was arbitrarily chosen. The mean
system efficiency (point C) is now reduced to the maintained system
efficiency (point D).

The remaining factor, the coefficient of utilization, is the
percentage of light which ultimately finds its way to the working
surface. In order to illustrate this principle, we have arbitrarily
chosen a depreciation factor of 0.65, which lowers the maintained
delivered system efficiency from point D to point E. The remainder
is useful light.

These factors are extremely important in enabling the lighting system owner to get the most for his money. Each of the factors
is highly variable: Each depends on the way the owner takes care
of his lighting system, on the size of light source he chooses, and
on the kind of luminaire he selects.

Figure 2-14 describes the characteristics of the various lamp
types (4). As can be seen, the incandescent lamp's principal problem is the low inherent efficiency of the light source. On the
other hand, the mercury lamp has large losses from point A to point
B. This occurs because a mercury lamp lasts far beyond its economic
life and remains in operation even though its light output has long
since deteriorated to a low value.

The fluorescent lamp shows a dramatic gain in useful light over
incandescent and mercury lamps. The introduction of new ballasts
that have high-efficiency steel cores, more efficient circuitry, and
more compact windings has reduced ballast losses (point B to point
C). However, the coefficient of utilization of fluorescent lighting would be poor if it were used to try to focus direct lighting
on a specific area.

The metal halide lamp shows a significant efficiency gain over
fluorescent lighting, but this gain depends on the actual use situation. For example, the typical 12-ft ceiling height of a store
would necessitate the use of relatively low-wattage metal halide
lamps, rather than 8-ft fluorescent lamps. However, with a 12-ft

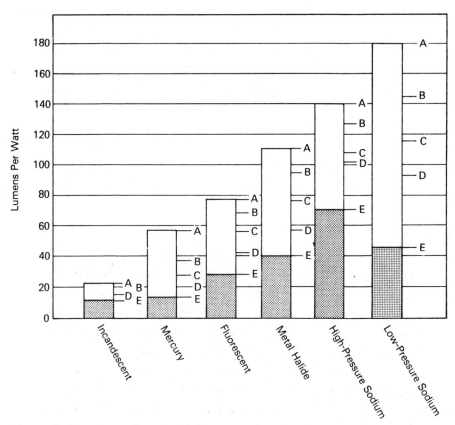

Figure 2-14. Lamp characteristics.

ceiling height, the useful available light from both lamp types would be nearly equivalent and could easily favor fluorescent lighting.

The higher pressure sodium lamp is characterized by high initial efficiency. Small losses due to dirt accumulation (between points C and D) are shown because modern luminaires with a very low luminaire dirt depreciation factor are available. In addition, the co-efficient of utilization (from D to E) may be somewhat better in current practice than shown, and this could result in still greater useful light.

The low-pressure sodium lamp is characterized by still higher
initial efficiency; however, the actual useful light may be less than
that of the high-pressure sodium lamp, as indicated at E^2. The principal reason is its low coefficient of utilization whenever light
must be directed into a limited area, such as a narrow roadway. For
example, in the case of its utilization at an underpass, the coefficient of utilization would be much higher and the resulting useful
light can be equal to or higher than that for high-pressure sodium
lighting, as indicated at E^1. Disadvantages are the hazards involved
in disposing of the used luminaires and its poor color rendition.

NONUNIFORM ILLUMINATION

The amount of light required for good task performance is dependent
on the amount of light needed for the tasks being performed. If just
the right amount is provided, and if the light source is in just the
right position above the activity, lighting energy can be kept to a
minimum. In this approach, luminaires are placed only where they
are needed, so that a nonuniform pattern usually results.

In the recent past it was virtually impossible to design an
effective nonuniform system because buildings were and still are
being built without builders knowing who the tenants are going to be,
what tasks will be performed, or where the work stations will be
located. Prior to today's new technology, the only way to help ensure the installation of adequate lighting was to distribute illumination uniformly in a space, in an amount generally sufficient
for the most demanding task likely to be performed in that space.
Although this practice tended to waste energy, there were no easy
alternatives.

Today there are. One of these is a flexible overhead ceiling
system. Luminaires fit into standard-size dropped ceiling modules
which are interconnected by means of branch wiring that can snap together and snap apart very quickly and easily. These luminaires
can be moved about to the work station areas and can provide the
appropriate amount of illumination needed for the tasks to be per-

formed. When the space is rearranged, the lighting system can be changed to fit the new pattern. Unfortunately, the limitations of the flexible overhead ceiling system include the high labor cost of making changes and the damage that is frequently done to the ceiling and lighting fixtures.

A second alternative to uniform lighting is to associate the lighting with the work station, so that lighting and work station can be moved together. In response to this need, a number of manufacturers of office furniture offer various systems that have lighting built into or attached to the furniture and, in some cases, involve portable direct/indirect lighting in the form of "kiosks."

Other approaches to nonuniform lighting include the following measures; either individually or in combination with others:

1. Installing switches to turn off unnecessary lights
2. Using lamps with variable light outputs required for specific tasks
3. Relocating or installing new fixtures to suit specific tasks
4. Controlling lamp intensity with dimmers
5. Using multilevel ballasts to obtain the proper light level required for a specific task

In order to select the most appropriate measures, one must make a careful analysis of task requirements, expected duration of the task to be performed, possible future relocation, the quality of illumination required for specific tasks, and the frequency with which lighting requirements change.

LIGHTING CONTROLS

Switching

The basic rationale for lighting controls is the provision of light where it is needed, when it is needed. Lighting systems must have control flexibility if energy is to be saved. Unfortunately, in recent years the number of switches and other controls devices for lighting systems have been minimized or omitted to save initial

costs. Thus, one switch may control whole sections, or even whole
floors of buildings, a situation which results in the waste of both
light and energy.

A 10 to 25% reduction in electric energy usage is possible with
better switching patterns; however, the controls should be human-
engineered to assure consistent use. Past studies have shown that
both the arrangement and labeling of switches are very important.
If a lighting panel is left unmarked, all of the lighting is likely
to be left on most of the time.

In addition, there is a need for sophisticated controls that
can be easily applied to a broad range of facilities from single-
story commercial buildings to multistory office structures. A number of such devices are now becoming available. The basic components of one such system are a central control device and a receiver/
switch which are located in the electrical supply line to the lighting fixture. This is a standard control arrangement, but it differs
from more traditional arrangements in that it can control small
loads individually, by using low-cost logic elements and digitally
coded signals for the functional and address commands to turn the
fixtures on and off. This can greatly increase system capacity,
since, by carefully defining the control hierarchy and digital word
structure, the potential number of remote control points is limitless.

A possible configuration of a system designed to control general lighting is shown in Fig. 2-15. A microprocessor carries out
a planned sequence of control operations according to a stored-
memory program. At the input end, commands may be generated by a
variety of devices: clock, photocell, touch pad, and so on. The
processed commands, properly coded, are then sent to the remote receivers, where they are decoded, causing the appropriate control
function to occur. In this particular case, the signal is shown
traveling between the transmitter and receivers by means of the
building's power wiring.

In operation, the system might utilize a "lighting map," or a
stored pattern of luminaires set to go on or off according to the

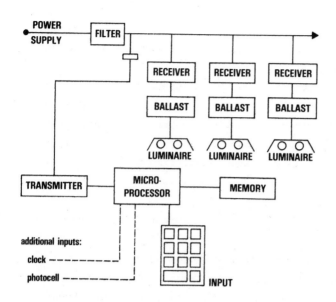

Figure 2-15. Lighting control system. (Data from General Electric.)

functional needs of the work space. During working hours a clock
would recall the appropriate patterns from memory, switching to different patterns as needed. One pattern might be: arrival--working--
lunch--working--departure--cleaning--nighttime. Local controls can
be used to override the normal pattern at any time, by injecting address and command codes for the particular luminaires involved. A
standard telephone may be used as an input control device simply by
setting up a telephone station as an input line and adding an inexpensive interface to the microprocessor.

Overall the use of stored programs, which can be accessed
through a variety of input devices to generate the desired control
action, greatly reduces the number of local control points and associated wiring without reducing flexibility. Changing lighting
patterns to suit new floor, partition, or work space arrangements
does not involve changing power wiring or moving switch legs; it
involves only a new set of easily changed memory instructions.

The key to the practicality of this particular control system
is the utilization of standard logic elements such as the micropro-

cessor. These devices are now being mass-produced in a variety of
configurations, and prices have dropped from hundreds of dollars per
unit a few years ago to a few dollars per unit today. Microproces-
sors themselves are fast becoming the ubiquitous heart of numerous
electrical and electronic appliances, from pocket calculators to
electric ranges and automobile ignition systems. One report estimates that U.S. shipments of "chip sets," which include micropro-
cessors and their auxiliary devices, will climb to some 10 million
units/year by 1980, from the 1971 level of 1000 and the 1974 level
of 200,000 units.

Two test installations are now utilizing this system and have
verified the approach. In the first installation, a single-room
office for two people, energy savings amounted to as much as 9
kWh/day, depending on whether both occupants were in the room and
on the length of time they occupied it. During a typical month,
lighting energy reductions averaged 47%. In this arrangement the
room lighting was arranged as it might be for one occupant in the
space.

The second test installation, which involves 400 fixtures in a
company headquarters building in New Jersey, has achieved lighting
energy reductions in excess of 50%. Local override capability, established via the telephone system, controls the lighting in several
types of office space. Each telephone acts as a light switch and
can control any fixture in the system.

Such control systems are not limited to lighting control; they
can perform other functions, since even with the most complex lighting system, the control microcomputer is idle much of the time.
Adding additional loads only increases the complexity of the control
program, not the system itself. Consequently, such a uniquely addressed switch approach can provide a practical and economic way
to control the hundreds of thousands of lighting and energy-using
loads in industrial and commercial spaces.

Dimmers

Dimming controls that save energy are available for almost all types of luminaires, including fluorescent and HID lamps. With the right wiring, individual luminaires can be dimmed to provide only the amount of lighting needed. Virtually all dimmers now on the market are of the solid-state type: electric power to the lamps is controlled by instantaneously limiting the time in which the sine wave power is applied. The lamps see a "chopped" or "phase controlled" power pulse, which limits energy use according to the setting of the dimmer control.

For discharge lamp systems, dimming reduces energy use in direct proportion to ilight output, with the lamp remaining relatively efficient. Incandescent lamps lose efficiency very rapidly as they are dimmed, so light output can be adjusted according to ambient light conditions. Its effect on lamp performance cannot be evaluated at this time.

Power Factor Correction Equipment

A high-power factor (voltage and current in "phase") on an electrical system means that power is being transmitted at maximum efficiency. Lowering the power factor increases line current without increasing power transfer, but energy losses rise as a result of line heating. Power factor correction is therefore desirable and can make an electrical system more efficient. From an economic standpoint, keeping the power factor of a building or industrial plant high will eliminate extra charges, or "power factor penalties," which utilities sometimes charge large users.

Demand Controllers

Utilities may also charge extra for a consumer's "demanding" more than his average energy use in a given time period. The demand controller is a combined sensing and switching device which dynamically

monitors the energy use of a facility and turns off preselected
electrical loads if demands reach critical levels. Demand controllers should be used to minimize demand charges.

Lighting Control Systems

Automated systems are being incorporated into many new and remodeled
buildings. Frequently, the heart of such a system is a large computer which is used to oversee a variety of building operations, including the HVAC system, fire warning systems, communications systems, security systems, and often, lighting systems. Operational
programs which act according to time-of-day or outside daylight conditions should be installed.

LUMINAIRES

Together the efficiencies of a lamp and its fixture determine the
quantity of light transmitted into a space for each watt consumed.
In choosing the luminaires or fixtures to be used, four basic ideas
should be kept in mind.

First, because the location of work stations and the tasks to be
    performed in those areas are subject to frequent changes, the
    lighting system should be flexible enough to permit relocation
    of luminaires quickly, easily, and at low cost.
Second, the lamp should be an efficient source of light, and as much
    light as possible should be able to get out of the fixture and
    onto the work surface.
Third, the fixture should not be a source of glare.
Fourth, wherever possible, the heat generated by lighting fixtures
    should be integrated into the heating, ventilating, and air-
    conditioning systems in order to minimize energy use.

There are two other factors that should be considered when
choosing a lighting fixture. The first is the finish of the fixture,
which should be suited to the environment of the space in which it
will be located. The second factor is the ease with which the fix-

ture can be cleaned. It should not be difficult to remove the diffuser to get at the lamps and inside surfaces of the fixture.

Since each person has his or her own idea as to what consti-
tues good appearance, fixture appearance can have an important impact on working conditions. For surface-mounted fixtures, a wide
variety of finishes and shielding materials is available, some of
which use wood to help soften the effect of too much plastic in an
enclosed space. For recessed fixtures, aesthetics can be enhanced
by installing the fixtures flush with the ceiling to provide a smooth,
finished appearance, as shown in Fig. 2-16.

Figure 2-16. Recessed fixtures. (Courtesy of General Electric.)

Another option is the use of regressed luminaires: the actual
luminaire is set above the ceiling plane, as indicated in Fig. 2-17.
These luminaires tend to disappear from the field of view as one
looks across the room and have the advantage of taking the workers'
attention off the overhead zone, thereby eliminating the feeling
that lighting dominates the room.

Maintenance

By maintaining a higher light output from existing fixtures, it may
be possible to reduce the wattage of the lamps in each fixture and,
in some cases, the number of fixtures in service, without a reduction in illumination. Three variables affect the system maintenance:

1. Lamp lumen depreciation (LLD), which is the drop in lamp lumen
    output over lamp life, due to the lamp characteristics
2. Luminaire (lighting fixtures) dirt depreciation (LDD), which is
    the drop in lumen output, due to accumulation of dirt on the in-
    terior reflective surfaces of the fixture and lens
3. Room surfaces, which begin to get dirty, reducing reflectivity.

The combination of these three variables results in an average reduction in light output of about 25% from fluorescent fixtures
(this will vary with room atmosphere and lamp types) over an ex-
tended period of time and despite the maintenance procedures that
are in general practice today.

The effect of these three factors is shown graphically in Fig.
2-18. To maximize cost-effectiveness of maintenance starting from
time zero, one needs to determine the optimum periodicity of maintenance intervals A, B, and C. At point 1 cleaning the lamps and
luminaires will restore 82% efficiency. In very dirty installations,
such as foundries, cleaning should be done every year or even every
6 months.

At point 2 it will pay to replace the lamps, even though they
are still burning. Let us call this interval B. Moreover, when a
maintenance crew goes through the area to replace lamps it should
also clean the luminaires at the same time. At point 3 it will pay

Table 2-17. Regressed fixtures. (Courtesy of General Electric.)

Table 2-7. Lighting Cost Comparison

	Lighting Method #1	Lighting Method #2
Installation Data		
Type of installation (office, industrial, etc.)		
Luminaires per row		
Number of rows		
Total luminaires		
Lamps per luminaire		
Lamp type		
Lumens per lamp		
Watts per luminaire (including accessories)		
Hours per start		
Annual burning hours		
Group relamping interval or rated life		
Light loss factor		
Coefficient of utilization		
Footcandles maintained		
Capital Expenses		
Net cost per luminaire		
Installation labor and wiring cost per luminaire		
Cost per luminaire (luminaire plus labor and wiring)		
Total cost of luminaires		
Assumed years of luminaire life		
Total cost per year of life		
Annual interest on investment		
Annual taxes		
Annual insurance		
Total annual capital expense		
Operating and Maintenance Expenses		
Energy expense		
Total watts		
Average cost per kWh		
Total annual energy cost		
Lamp renewal expense		
Net cost per lamp		
Labor cost each individual relamp		
Labor cost each group relamp		
Percent lamps that fail before group relamp		
Renewal cost per lamp socket per year**		
Total number of lamps		
Total lamp renewal expense per year		

Cleaning expense
  Number of annual washings _____
  Man-hours each (est.) _____
  Man-hours for washing _____
  Number of annual dustings _____
  Man-hours per dusting each _____
  Man-hours for dustings _____
  Total man-hours _____
  Expense per man-hour _____
  Total annual cleaning expense _____
Repair expenses
  Repairs (based on experience, repairman's time, etc.) _____
  Estimated total annual repair expense _____
  Total annual operating and maintenance expense _____
Recapitulation
  Total annual capital expense _____
  Total annual operating and maintenance expense _____
  Total annual lighting expense _____

*Total annual energy cost $= \dfrac{\text{Total watts} \times \text{burning hours per year} \times \text{cost per kWh}}{1000}$

**The following formulas give the annual cost per socket for lamps and replacement. They also can be used to determine the most economical replacement method.

Individual replacement $= \dfrac{B}{R}(c + i)$ dollars/socket/year.

Group replacement $= \dfrac{B}{A}(c + g + KL + Ki)$ dollars/socket/year.
(early burnouts replaced)

Group replacement $= \dfrac{B}{A}(c + g)$ dollars/socket/year.
(no replacement of early burnouts)

where B = burning hours per year
R = rated average lamp life, hours
A = burning time between replacements, hours
c = net cost of lamps, dollars
i = cost per lamp for replacing lamps individually, dollars
g = cost per lamp for replacing lamps in a group, dollars
K = proportion of lamps failing before group replacement (from mortality curve)
L = net cost of replacement lamps, dollars

No general rule can be given for the use of group replacements; each installation should be considered separately. In general, group replacement should be given consideration when individual replacement cost i is greater than half the lamp cost c and when group replacement cost g is small compared to i.

Source: IES Lighting Handbook, 1981 Reference Volume.

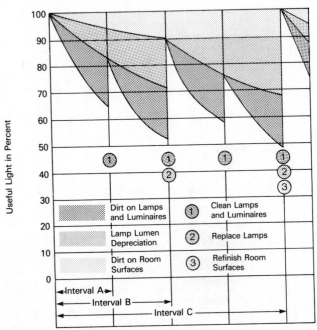

Figure 2-18. Systematic lighting maintenance. (Derived from data in the IES Lighting Handbook, 1981 Application Volume.)

to perform all three functions, including refinishing the room sur-face. Table 2-7 can be used to make a cost comparison of various lighting methods and to determine the most economical replacement method.

Figure 2-19 shows how light output of luminaires in three typi-cal situations diminishes as dirt collects on lamps and luminaire surfaces. In extremely dirty environments, such as those in found-ries, dirt depreciation can reduce the light output of a lighting system to less than 50% of the initial light output in only 24 months of system operation. This means that if one were to pay $1000 for a new lighting system and then not maintain it, one would soon be

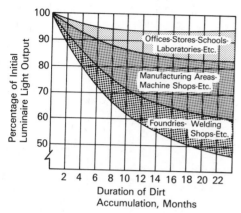

Figure 2-19. Luminaire depreciation. (Derived from data in the IES Lighting Handbook, 1981 Reference Volume.)

getting only *half* his money's worth, but still be paying money and using energy at the full rate. Put another way, one could, instead, buy $500 worth of light just by cleaning and putting in new lamps.

BALLASTS

A new 35-W, 4-ft lamp, which is electrically matched to a new high-efficiency ballast, has recently been developed. Called the high-performance system, the new ballast is characterized by high-efficiency steel, which is used for the core of the transformer; by winding coils with flattened wire, which permits greater transmission of internal heat; by improved circuitry; and by non-PCB capacitors for environmental compatibility. These improvements cut ballast losses in half by allowing the ballasts to run at substantially lower operating temperatures; which results in extending the life of the ballast twice its normal life. Also available to the public are reduced output ballasts, including low-energy types, which operate lamps at lower than normal current, and high/low ballasts, which offer the option of two levels of light output.

Table 2-8 compares the high-performance system, which uses various fixtures to the standard 40-W, 4-ft system. The implications

Table 2-8. *High-Performance Lighting Systems and Other Fixtures**

HIGH-PERFORMANCE LIGHTING SYSTEMS

4-Lamp Troffer (Enclosed) Fixture Comparison:

	High Performance System	Standard vs. F40 System
Relative Light Output	104%	100%
Watts (2 Lamps + Ballast)	74	88
Watts Saved	16%	—
Light Output Per Watt Increase	23.5%	—

OTHER FIXTURES

4-Lamp Wrap-Around Surface Mounted

	High Performance vs. System	Standard F40 System
Relative Light Output	106.5%	100%
Watts (2 Lamps + Ballast)	71	82.5
Watts Saved	14%	—
Light Output Per Watt Increase	24%	—

2-Lamp Strip Surface Mounted

	High Performance vs. System	Standard F40 System
Relative Light Output	100%	100%
Watts (2 Lamps + Ballast)	78	95
Watts Saved	18%	—
Light Output Per Watt Increase	22%	—

2-Lamp Strip Industrial Vented

	High Performance vs. System	Standard F40 System
Relative Light Output	97.5%	100%
Watts (2 Lamps + Ballast)	78	96
Watts Saved	19%	—
Light Output Per Watt Increase	20%	—

*Values shown are relative only with systems in the same fixture. Comparison between fixture types is not valid. All fixture tests conducted at 77°F ambient.

Source: Data from General Electric.

for energy conservation are apparent, since energy savings of 14 to 19% over conventional systems are possible.

It is clear then that for new installations and relighting, the selection of the high-performance system is recommended over the standard system. For existing installations, the substituion of new high-efficiency 35-W lamps for the present 40-W, 4-ft lamps is also recommended on any type of rapid-start ballast. Moreover, as standard ballasts fail in existing installations, they should be replaced by the new high-performance ballasts. A hypothetical example should underscore the reason: Each ordinary ballast is an installation of 1000 two-lamp, 40-W fluorescent luminaires consumes 12 to 16 W, which amounts to an annual energy consumption of 24,000 to 32,000 kWh for a building operating 2000 h per year. In comparison, an efficient ballast that has a 10-W loss will use 20,000 kWh annually, for a savings of 4000 to 12,000 kWh per year. Therefore, the substitution of high-performance ballasts for standard ballasts saves on energy costs as well as labor costs in the long run.

DAYLIGHTING

Properly controlled natural light can reduce the amount of artificial lighting required in a building, thereby saving energy. The design problem is to transform the direct rays, which usually cause glare and excessive footcandle levels, into a softer, less bright, more useful light that can be used for total illumination or as a supplement to artificial light.

Daylighting does not reduce the amount of artificial light that must be installed in a building. Rather, it reduces the total amount of energy consumed by lighting in those areas which receive sufficient natural light to allow the lights to be turned off for part of the occupied hours. Daylighting requires proportionately more glazing (glass and/or windows) in the building envelope, which usually increases heat transmission and, possibly, infiltration. In cold weather, heat transmission (heat loss) and infiltration of cold air into the building, plus the absence of interior heat that is a by-

product of artificial light, will increase heating loads, but solar
heat gains can compensate for this relatively small portion of the
total heating load. In hot weather, however, the infiltration of
hot air into the building, transmission loads, and direct solar gains
will increase demands on the air conditioning system.

There are no simple solutions to the problem of balance between
daylighting and electric lighting, and all that can be offered at
this point are general guidelines. A careful cost/benefit analysis
should be conducted to determine the optimal balance for the system. Heat loss and a northern exposure, the biggest problems in
northern climates, can be offset by covering windows at night with
draperies or venetian blinds. In some instances it may be practical to reduce the window area. In southern climates the biggest
problem is heat gain. Since heat gain is worst in windows with
southern and western exposures, these windows should be covered so
that direct sunlight is not admitted. This also helps to control
glare. It should be remembered that once sunlight is transmitted
through the glass, even though interior blinds are used, energy becomes heat on the inside of the building. Thus, it may be practical to add shading devices on the outside of the building to prevent
sunlight from reaching the glass. Another solution is to replace
the glass with high-performance glass, that is, a double- or triple-
glazed window, which helps to control both heat gain and heat loss.

Skylights should also be considered as a means of providing
daylight. Various studies have indicated that if electric lighting
were to be turned off when daylight is available through skylights,
as much as 40 W of electric power per square foot of skylight area
could be saved.

But here again, in order to determine the net yearly increase
or reduction in energy consumption, the amount of energy saved by
using skylights must be analyzed in conjunction with the increased
amount of energy required for heating and the increase or reduction
in heat gain and cooling load.

The use of skylights has long been subject to criticism for the
quantity of heat transmitted. This criticism is often valid, since

Table 2-9. *Comparative Costs of Skylights, Fluorescent Lights, and the Two in Combination*

Cost Consideration	Source of Illumination		
	Skylights[a]	Fluorescent Light Units[b]	Skylights and Fluorescent Units[c]
First Cost of Lighting Installation (Less Lamps)	$510.00	$720.00	$950.00
Uniform Annual Cost of Recovering First Cost at 4% for an Operating Period of 25 Years	$ 20.40	$ 28.80	$ 38.00
Annual Cost for Insurance at 1.2% of First Cost	$ 6.12	$ 8.64	$ 11.40
Annual Cost for Lamps[d]	––	$ 18.00	$ 11.00
Annual Labor Costs for Cleaning and Relamping[e]	$ 15.00	$ 40.00	$ 39.42
Annual Power Cost[f]	––	$173.88	$106.26
Total Annual Lighting Cost	$ 41.52	$269.32	$206.08
Annual Cost Per Footcandle	$ 0.83	$ 5.43	$ 4.12

Note: Costs are based on figures supplied by reputable manufacturers and are believed to be typical. They should be treated, however, as rough approximations only for preliminary comparisons. Consideration should be given to local changes of sky conditions.

[a]Six skylights, 37 × 37 in. inside dimension, double dome of high-transmission acrylic, 10 ft. on centers, 9-ft room ceiling, average lighting level under a 5000 fc uniform sky - 50 fc

[b]Eighteen fluorescent units, two 40-W lamps, 45° louvers, 9-ft room ceiling, 30 × 30 ft room, average lighting level = 50 fc.

[c]Six skylights and 11 fluorescent units, average lighting level = 50 fc.

[d]All lamps replaced once every 3 years.

[e]Cleaned every 18 months.

[f]Based on $0.035/kWh, 3000 hr per year.

Source: National Research Council, *Solar Radiation Considerations in Building Planning and Design*, National Academy of Sciences, 1976, p. 155.

in very general terms it may be said that light and heat are essentially the same thing, so that where there is light, there is
also heat—almost in direct proportion. However, skylights produce
less heat per unit of light than most equivalent electric lighting
systems. Table 2-9 shows that daylighting produces more lumens per
watt than electric lighting of the more common varieties and, in
terms of air conditioning needs, that daylighting requires less cooling per unit of light.

We should also consider the costs of lighting systems, whether
they are electric or not. Table 2-10 presents a typical comparative
cost analysis for three different systems of lighting—skylights
only (in this case, plastic-domed skylights), fluorescent lights
only, and skylights and fluorescent lights together. In this analysis, all normal expenditures directly related to the various lighting systems were combined. Notice that daylight is the least expensive method of lighting.

Of course, factors other then economics are involved. There is
the visual performance level to be considered, as well as the general atmosphere being created by the architect, both of which, in
the final analysis, are the primary decision factors.

Table 2-10. Efficiency Comparison of Skylights and Electric Lighting

Source of Illumination	Light Output		Air Conditioning Load (Tons/100,000 Lumens)
	Lumens/Ft^2	Lumens/Watt	
Daylight Through High-Transmission Acrylic Plastic	5,270	106	0.27
Daylight Through Medium-Transmission Acrylic Plastic	3,110	106	0.27
Incandescent Light	——	20	1.90
Fluorescent Light	——	60	0.63

Source: National Research Council, *Solar Radiation Considerations in
Building Planning and Design,* National Academy of Sciences, 1976,
p. 154.

TOTAL BUILDING THERMAL OPERATION

Use of the most efficient sources is always recommended for lighting systems, but whenever changes are made in lighting efficiency of hours of operation, the effect on heating and cooling must also be assessed to obtain the net impact on energy use. Total building thermal operation involves the interaction of many subsystems, primarily the relationship between lighting, heating, and cooling. Handling this relationship in an effective manner means installing lighting that can help in heating the building and, conversely, can have a minimal impact on attempts to cool the building. This latter point is particularly important.

Today's technology can bring lighting heat under control and can lower ceiling temperature by several degrees in the summer by drawing off lighting heat to prevent a buildup above the ceiling. The resultant reduction in radiant temperature is a vital factor in maintaining worker comfort, as thermostats are set at 78°F.

The heat for lighting systems operated by coal-, hydro-, or nuclear-generated electric energy can make an important contribution to reducing oil and gas consumption in winter. Unfortunately, this is a point not yet appreciated by many. For example, when one state closed its schools and lowered the thermostats during a recent winter gas shortage, this "conservation" measure reduced coal usage (which would have generated the electric energy for the lighting that was not used), but slightly increased gas use. Heat generated from the lights and from the students would have a higher temperature rise than the 15° reduction in thermostat setting. This kind of energy action should be put into effect with full knowledge of all the ensuing consequences, or the results may be counterproductive.

The approach to conserving energy in buildings should be based on the contribution of each subsystem to the effective operation of the building. Internal transportation, heating, cooling, process energy, and lighting should all use the smallest amount of energy possible, consistent with enabling people to do their jobs. Lighting as a subsystem is one of the most obvious energy users and has, in some cases, led to the simplistic conclusion that reducing lighting is a major energy conservation measure.

Lighting accounts for only 5% of the country's total energy consumption and uses the less critical energy sources (coal, hydro power, and nuclear power). Curtailment of all interior lighting would increase space heating by 2% (in far more critical fuels--oil and gas) and save only ½% in cooling energy. This suggests that the net effect of lighting on energy use may not be 5%, but only 3½% (2% heating + ½% cooling = 3½% net energy).

CONCLUSION

One of the tragedies of the past has been the tendency to continue poor lighting practices--largely because it was easier and could be accomplished without any greater capital investment. Both the lighting industry and its critics seem to agree that lighting equipment is frequently allowed to operate even when it is no longer needed.

The current inventory of buildings has, in general, rather limited opportunities for controlled usage of lighting. This is primarily the result of two factors, namely, the *modular nature of building design* and the *first-cost syndrome*. Overcoming this situation so as to permit good lighting when and where it is needed and still reduce unnecessary usage certainly seems worthy of attention.

The Illuminating Engineering Society's 12 recommendations for saving energy in illumination systems, without sacrificing visual performance and visual requirements, provides a good summary of effective lighting practices. The recommendations are as follows:

1. *Design lighting for expected activity.* It is necessary to determine what types of activities are expected, their possible duration, and where they will occur. Lighting should be provided for the seeing tasks; less light is required for the surrounding nonworking areas, such as corridors, and storage and circulation areas. There should be a capability for relocating and altering lighting equipment whenever changes in use of the space are anticipated.

2. *Design with more effective luminaires and fenestration.*
Lighting equipment and windows capable of providing proper visibility
for the performance of tasks should be selected. This includes selecting lighting equipment designed to avoid ceiling reflections and
glare, windows with controls for sun, and skylights.

3. *Use efficient light source.* Since different lamps have different properties, the choice should be the most efficient source
that is appropriate to the app ications. Light source color, life,
and physical size are also characteristics to be considered.

4. *Use more efficient luminaires.* More efficient luminaires
produce a greater amount of light for the wattage consumed; however, consideration should also be given to ceiling reflections and
glare and to the ease of cleaning and relamping.

5. *Use thermally controlled luminaires.* Recognize lighting
heat. Heating and cooling systems should be designed to use lighting heat for reducing heating energy and to control lighting heat
for minimal effect on cooling.

6. *Use care in choosing the finish on ceilings, walls, floors,
and furnishings.* Dark finishes absorb light, while very light finishes can cause glare. One should follow IES reflectance recommendations.

7. *Use efficient incandescent lamps.* Higher wattage incandescent lamps are more efficient. Thus, using fewer higher wattage
lamps can save power. The less efficient long-life lamps consume
more energy for the same light output than general service lamps.

8. *Provide flexibility in the control of lighting.* Separate
and convenient switching or dimming devices should be used for areas
that have different activity patterns. Photoelectric control of the
electric lighting may be considered when adequate daylighting is
possible. Institute a program that will remind occupants to turn
off the lights as they leave an empty room or when daylighting is
adequate.

9. *Design fenestration to control heat-producing radiation entering a space.* This will reduce the cooling load and still make
use of available daylight.

10. *Design fenestration to use daylighting as practicable to produce the required illumination, either alone or with an electric lighting system.*

11. *Select luminaires with good cleaning capability and lamps with good lumen maintenance.* The appropriate lighting servicing plan should be selected to minimize light loss during operation, thereby reducing the number of luminaires required. Lighting equipment should be kept clean and in good working order by using a well-planned program of regular cleaning, relamping, and servicing.

12. *Post instructions covering operation and maintenance.* These should initially be based on the design criteria but may be modified later as more efficient, newer replacement equipment is installed. As activity locations change, the lighting system and instructions should be modified accordingly.

Optimizing our lighting investment involves two basic ideas: (1) buying the lighting system that gives us the most for our lighting dollar, and (2) ensuring that it continues to give us the light we originally purchased and continue to pay for in our electric bill. To determine which lighting system will give us the most for our lighting dollar, *life-cycle costing techniques* should be used. Life-cycle costing is regarded as a truer measure of the impact of a system on both economics and the environment than any other yardstick. It considers the total dollars spent on buying, installing, operating, and maintaining a piece of equipment or a system during its lifetime.

Given the opportunity through equipment capability and education, building owners and operators should be motivated by the cost of electricity to appropriately control the use of their lighting system. At the same time, they can have good lighting and still contribute to the conservation of the nation's energy resources.

NOTES AND REFERENCES

1. Studies show lighting affects productivity, Elec. World, vol.
    183, no. 12, June 15, 1975, pp. 86-87.

2. Federal Energy Administration, Check reading and verification ex-
    periment, *Lighting and Thermal Operations: Energy Conservation
    Principles Applied to Office Lighting,* Conservation Paper no. 18,
    Apr. 15, 1975, pp.   -23.

3. Federal Energy Administration, *Guidelines for Saving Energy in
    Existing Buildings: Building Owners and Operators Manual, ECM 1,*
    Conservation Paper no. 20, June 16, 1975, pp. 239-241.

4. Figure 2-14 must not be used for decision-making; it is intended
    only to indicate how lighting variables may operate.

3

Improving Appliance Efficiency

Melvin H. Chiogioji
U.S. Department of Energy
Washington, D.C.

American homes are filled with numerous appliances that operate at varying degrees of efficiency, depending on appliance design and use. The amount of energy consumed by these appliances is primarily a function of the extent to which they are used; but it is also affected by the ways in which they are used and maintained.

Appliance energy usage accounts for about 25 to 30% of the total energy consumed by an average home (see Fig. 3-1). The calculations in Fig. 3-1 are based on aggregate consumption; thus, they do not reflect the interactions that occur between appliance usage and heating and cooling needs.

Many things can be done to reduce the amount of energy used by appliances. Obviously, one of the first steps is to reduce usage of the appliances, and those that account for the largest amount of energy consumption deserve the most attention. Table 3-1 lists commonly used electrical appliances and the average amount of energy they consume each year.

But averages do not necessarily tell the whole tale. In fact, appliances operate within a very broad range of efficiency. Tables 3-2 and 3-3 provide summaries of the highest and lowest efficiency

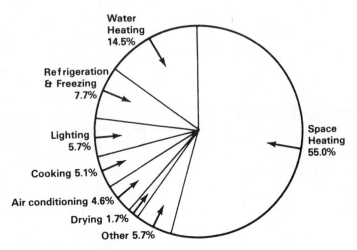

Figure 3-1. Residential energy use in the United States. Residential primary energy consumption, 1970 (13 QBtu). (From National Science Foundation Report R-1641-NSF, June 1975, p. vi.

units of major electric and gas appliances. Operating the appliances at the efficiencies listed in Tables 3-2 and 3-3 will result in the annual operating costs shown in Table 3-4. As is readily evident, substantial cost savings can be achieved simply through buying the most energy-efficient models currently available on the market.

A major factor in appliance energy conservation is the user. In fact, a recent study in which homemakers prepared identical meals on identical equipment indicated that some used only half as much energy as others, because of energy-wise practices (1). Whether home appliances are new or old, they can be operated in ways that will keep the operating costs down. Moreover, appliances which are used efficiently last longer and give more trouble-free operation.

The remainder of this chapter discusses the major appliances, used in the home and the newer technologies that are available for enhancing energy efficiency.

HEATING AND COOLING EQUIPMENT

Many improvements in the efficiency of modern heating and cooling
equipment have been made. Since most heating and cooling systems
will use many times their original purchase price in fuel, a small
improvement in efficiency can mean a large savings in fuel cost over
the years.

Some of the more important considerations in trying to make the
appropriate individual choice of heating and cooling systems are
regional weather characteristics, type of house, initial cost, and
cost of operation. The size of the system is determined by the demands on the system and the efficiency of its components. Demands
on the mechanical system are extremely variable; they depend on the
natural conditions outside the home, the insulating effect of the
building's structure, and the comfort zone parameters within the
home. The relationship among these factors must be carefully balanced to provide thermal comfort while maximizing energy efficiency.

Considerable controversy exists over the typical operating efficiencies of these systems. One reason is the inconsistencies in
the definition of the term *efficiency*. Until recently, the most
common efficiency measure was the steady-state, or full-load, combustion efficiency, which is defined as the ratio of useful heat
delivered to the furnace bonnet divided by the heating value of the
fuel. However, this definition does not give a complete measure of
the amount of fuel required to heat a living space throughout the
heating season. A definition that does, and one that is increasingly being used, is the seasonal performance factor (SPF), or
seasonal efficiency. This measure is defined as the ratio of the
useful heat delivered to the home to the heating content of the fuel
used by the furnace throughout the entire heating season. Seasonal
efficiency accounts for all of the factors affecting the heating
system's performance in its actual operating environment.

Table 3-1.   Annual Energy Requirements of Electric Household Appliances.

	Average Wattage	Est. kwh Consumed Annually
Health & Beauty		
Germicidal Lamp	20	141
Hair Dryer	600	25
Heat Lamp (Infrared)	250	13
Shaver	15	0.5
Sun Lamp	279	16
Tooth Brush	1.1	1.0
Vibrator	40	2
Home Entertainment		
Radio	71	86
Radio/Record Player	109	109
Television		
Black & White		
Tube Type	100	220
Solid State	45	100
Color		
Tube Type	240	528
Solid State	145	320
Housewares		
Clock	2	17
Floor Polisher	305	15
Sewing Machine	75	11
Vacuum Cleaner	630	46

	Average Wattage	Est. kwh Consumed Annually
Food Preparation		
Blender		
Broiler		
Carving Knife		
Coffee Maker		
Deep Fryer		
Dishwasher		
Egg Cooker		
Frying Pan		
Hot Plate		
Mixer		
Oven, Microwave (only)		
Range		
With Oven		
With Self-Cleaning Oven		
Roaster		
Sandwich Grill		
Toaster		
Trash Compactor		
Waffle Iron		
Waste Dispenser		
Food Preservation		
Freezer		
Manual Defrost		
Automatic Defrost		

*Based on 1000 hr of operation per year. This figure will vary widely depending on area and specific size of unit. See EEI-Pub #76-2 *Air Conditioning Usage Study* for an estimate for your location.

Source: *Annual Energy Requirements of Electric Appliances,* Edison Electric Institute, New York, 1975.

Average Wattage	Est. kwh Consumed Annually		Average Wattage	Est. kwh Consumed Annually
		Refrigerators/Freezers		
300	1	Manual Defrost		
1,140	85	12.5 cu. ft.	—	1,500
92	8	Automatic Defrost		
1,200	140	— 17.5 cu. ft.	2,250	
1,448	83	Laundry		
1,201	363			
516	14	Clothes Dryer	4,856	993
1,196	100	Iron (Hand)	1,100	60
1,200	90	Washing Machine		
127	2	(Automatic)	512	103
1,450	190	Washing Machine		
		(Non-Automatic)	286	76
12,200	700	Water Heater	2,475	4,219
12,200	730	(Quick-Recovery)	4,474	4,4811
1,333	60	Comfort Conditioning		
1,161	33	Air Cleaner	50	216
1,146	39	Air Conditioner (Room)	860	860*
400	50	Bed Covering	177	147
1,200	20	Dehumidifier	257	377
445	7	Fan (Attic)	370	291
		Fan (Circulating)	88	43
		Fan (Rollaway)	171	138
16 cu. ft.	1,190	Fan (Window)	200	170
16.5 cu. ft.	1,820	Furnace Fan	500	650
		Heater (Portable)	1,322	176
		Heating Pad	65	10
		Humidifier	177	163

Table 3-2. Estimate of Average Annual Energy Consumption of Electric Appliances

Product Type	Average Annual Energy Consumption		Percentage of Households Having Appliance	Remarks
	High Efficiency Unit	Low Efficiency Unit		
Refrigerator	504 kWh	780 kWh	9.7	14 Cubic Feet, Manual Defrost
Refrigerator-Freezer	1008 kWh	1908 kWh	95.6	17 Cubic Feet, Automatic Defrost, Top Mounted Freezer
Freezer	1176 kWh	2028 kWh	45	20 Cubic Feet, High Efficiency Manual Defrost, Chest; and 20 Cubic Feet, Low Efficiency Automatic Defrost, Upright
Dishwasher	1240 kWh	1780 kWh	42	Standard Size, 416 Cycles per Year, Electric Water Heater
Clothes Dryer	880 kWh	960 kWh	43	Large Size, 416 Cycles per Year
Water Heater	6020 kWh	7400 kWh	40	52 Gallon Capacity, 64.3 Gallons per Day
Room Air Conditioner	645 kWh	1400 kWh	55	10,000 BTUs per Hour, 750 Compressor Hours per Year
Home Heating Equipment	1600 kWh	1600 kWh		1,000 Watt Heater, 1600 Hours per Year

Television (Color)	170 kWh	366 kWh	85	19 Inch, 2200 Hours per Year
Television (Black and White)	107 kWh	150 kWh	99	19 Inch, 2200 Hours per Year
Conventional Range	675 kWh	825 kWh	71	30 Inch Oven
Clothes Washer	1250 kWh	1440 kWh	75	Standard Size, 16 Cycles per Year, Electric Water Heater
Dehumidifier	485 kWh	618 kWh	18	20 Pints per Day, 1300 Hours per Year
Central Air Conditioner	3000 kWh	5769 kWh	26	30,000 BTUs per Hour, 1,000 Compressor Hours per Year
Heat Pump	9395 kWh	14,655 kWh	3	*50,000 BTUs per Hour, 2080 Heating Load Hours per Year. (Note: Heat Pump Efficiency Varies with Local Climate and Dwelling Design)
Furnace	23,360 kWh	23360 kWh	14	*50,000 BTUs per Hour, 2080 Heating Load Hours per Year
Household Lighting	900 kWh	1100 kWh	100	High Efficiency Estimate Based on Increased Use of Flourescent and High Efficiency Incandescent Bulbs

*Design heating requirement of the house.

Source: *Annual Energy Requirements of Electric Appliances*, Edison Electric Institute, New York, 1975.

Table 3-3. Estimate of Average Annual Energy Consumption of Gas
Appliances

	Average Annual Energy Consumption	
Product Type	**High Efficiency Unit**	**Low Efficiency Unit**
Clothes Dryer	45 Therms	65 Therms
Water Heater	345 Therms	460 Therms
Conventional Range	60 Therms	130 Therms
Furnace	1003 Therms	1360 Therms

Note: Energy costs can be determined by multiplying data on energy
consumption by gas utility rates. One therm equals 100,000 Btu.

*Design heating requirement of the house.

Source: Annual Energy Requirements of Electric Appliances, Edison
Electric Institute, New York, 1975.

Another difficulty in assessing a system's operating efficiency
is the fact that remarkably little is known about system performance
in actual operating environments. The literature in this area is
replete with inconsistent information. With the above in mind, this
section attempts to describe a number of operating procedures and
new design technologies which can improve the energy efficiency of
heating and cooling systems.

Space Heating Systems

While new devices are continually being developed, furnaces with improved efficiency levels are now available. Table 3-5 gives a breakdown of the eight most important heating systems in existing single-
family residences. As is evident, the gas central forced-air system is the most common heating system, but one should particularly
note the dramatic percentage increase in the market share for elec-

Percentage of Households Having Appliance	Remarks
15	Large Size, 416 Cycles per Year
55	40 Gallon Capacity, 64.3 Gallons per Day
29	30 Inch Oven
44	*50,000 BTUs per Hour, 2080 Heating Load Hours per Year

tric central forced-air systems--a jump from 0.9 to 3.2%, a 255% increase. If electric heating continues its rapid growth, by 1980 it could bypass oil as the second leading heating fuel in the residential sector.

The percentage of central forced-air gas furnaces to the total number of gas heating systems in single-family residences increased from 58% in 1970 to 64% in 1975. During that same period, the percentage of central forced-air oil furnaces grew from 44 to 51% of the total number of oil systems in existing single-family reisdences. The electric heat pump has made only modest gains.

Table 3-6 provides estimates of the energy and cost (both purchasing and operating) impacts of several energy-conserving options for residential gas, oil, and electric space heating systems. The options include not only design changes in current technology, but also new technologies that might be introduced in the market within the next 5 years (such as the gas heat pump). The energy and cost impact estimates were developed to help evaluate the energy conservation potential and cost-effectiveness of design changes which were implemented by the manufacturer or installer of new heating equipment.

Table 3-4. *Estimate of Appliance Average Annual Operating
Costs*

Product Type	Average Annual Operating Costs	
	High Efficiency Unit	Low Efficiency Unit
Refrigerator	$ 25.	$ 39.
Refrigerator-Freezer	50.	95.
Freezer	58.	101.
Dishwasher	69.	87.
Water Heater (Gas)	126.	168.
Water Heater (Electric)	300.	368.
Clothes Dryer (Gas)	16.	24.
Clothes Dryer (Electric)	44.	48.
Room Air Conditioner	32.	70.
Central Air Conditioner	149.	287.
Heat Pump	467.	728.
Television (Color)	8.	18.
Television (Black and White)	5.	7.
Clothes Washer	62.	71.
Furnace (Gas)	390	488.
Furnace (Oil)	582.	709.
Furnace (Electric)	1,152.	1,152.
Dehumidifier	31.	24.
Conventional Range (Gas)	23.	53.
Conventional Range (Electric)	34.	41.
Microwave Oven	4.	7.

Note: Energy costs are based on 1980 estimated averages of
4.97¢/kWh, 36.7¢/therm of gas, and 84.0¢/gal of oil.

*Design heating requirement of the house.

Source: Annual Energy Requirements of Electric Appliances,
Edison Electric Institute, New York, 1975.

Remarks
14 Cubic Feet, Manual Defrost
17 Cubic Feet, Automatic Defrost, Top Mounted Freezer
20 Cubic Feet, High Efficiency, Manual Defrost, Chest 20 Cubic Feet, Low Efficiency, Automatic Defrost, Upright
Standard Size; 416 Cycles per Year
40 Gallon Capacity, 64.3 Gallons per Day
52 Gallon Capacity, 64.3 Gallons per Day
Large Size, 416 Cycles per Year
Large Size, 416 Cycles per Year
10,000 BTUs per Hour, 750 Compressor Hours
30,000 BTUs per Hour, 1000 Compressor Hours
*50,000 BTUs per Hour, 2080 Heating Load Hours per Year. *(Note: Heat Pump Efficiency Varies with Local Climate and Dwelling Design)
19 Inch, 2200 Hours per Year
19 Inch, 2200 Hours per Year
Standard Size, 416 Cycles per Year
*50,000 BTUs per Hour, 2080 Heating Load Hours per Year
*50,000 BTUs per Hour, 2080 Heating Load Hours per Year
*50,000 BTUs per Hour, 2080 Heating Load Hours per Year
20 Pints per Day, 1300 Hours per Year
30 Inch Oven
30 Inch Oven
600 Watts

Table 3-5. Eight Popular Heating Systems in Existing Singe-Family Homes (Percentage)

System	Year				% Change from 1970 to 1975
	1970	1973	1974	1975	
Gas Central Forced-Air	33.0	36.5	32.2	32.8	−1
Oil Central Forced-Air	10.8	10.9	12.0	11.8	9
Oil Boiler	7.7	7.9	7.5	7.4	−4
Gas Floor or Wall Furnace	7.9	8.7	7.2	7.0	−11
Built-In Electric	4.4	4.9	6.4	6.6	50
Electric Central Forced-Air Furnace[b]	0.9	2.0	2.6	3.2	255
Electric Heat Pump[b]	0.8	1.1	1.2	1.3	63
Gas Boiler	5.0	4.6	4.2	4.0	−20
Other	29.5	23.4	26.7	25.9	−12

Source: Dennis O'Neil, *Energy and Cost Analysis of Residential Heating Systems,* ORNL/CON-25, Oak Ridge National Laboratory, Oak Ridge, Tenn., July 1978, p. 16.

Properly Sizing the Unit. In the past, it was common practice to oversize heating systems, sometimes by as much as a factor of 2. Properly sizing the furnace to the actual thermal comfort require- ments can result in a savings of both initial capital costs and operating costs.

Adjusting the Bonnet Thermostat to $5°F$ Above Room Temperature. A furnace has a thermostat that can sense bonnet air temperatures and can control the temperature at which the circulating air fan turns on and off. Typically, the fan is set to turn on at temperatures between 125 and $140°F$ and turn off at $100°F$. Thus, most of the heat stored in the heat exchanger, after the blower stops, is not dis- tributed to the heated space and is lost. Adjusting the thermostat so that the fan turns on and off at $5°F$ above room temperature can substantially reduce annual energy use and cost.

Installing an Automatic Flue Damper. During the off cycle, heat escapes up the flue of the standard atmospheric combustion furnace. To prevent this heat loss, an automatic flue damper, which closes the flue during the off cycle, can be installed. Care should be taken in the installation of these systems in order to avoid safety problems.

Switching to a Sealed Combustion Unit. A sealed combustion furnace has its combustion chamber separated from the conditioned air in the house. Its combustion air is drawn directly from the outside and in many cases is preheated by the hot combustion products in the flue. A sealed combustion unit is more efficient than a comparable atmospheric combustion unit because it does not use the conditioned air, which was heated by the furnace, for combustion.

Reduction of Burner Nozzle Size. One of the most effective and easily implemented retrofit actions for oil furnaces involves the reduction of burner fuel nozzle size. Both steady-state and cycle efficiencies will improve as a result of reducing the firing rate, and a smaller nozzle size will enable the burner to operate for a larger fraction of the time. Recent field data demonstrate that average reductions in firing rates of 25% can reduce fuel consumption by 7%.

Burner Steady-State Efficiency Adjustment. The purpose of the burner equipment is to mix the fuel and combustion air in the proper proportions for efficient combustion. The steady-state efficiency of the burner-heat transfer equipment is a function of the percentage of CO_2 in the flue gases and the net stack temperature, with the most efficient operation corresponding to the maximum percentage of CO_2 and the minimum stack temperature. It has been estimated that on a national basis an average increase of 2% in efficiency can be achieved by adjustment of the steady-state efficiency.

Retention-Head Burner with Reduced Firing Rate. The installation of a retention-head burner could substantially increase the steady-state efficiency of a conventional oil heating system by reducing excess air requirements. And since the selection of a burner designed for operation at low firing rates (less than 0.75 gal/hr) ensures optimum performance, the combination of a retention-head burner

Table 3-6. Heating Systems Energy Savings with Various Design Changes

Option	Gas Systems		
	% Energy Savings	% Cost Increase	Nominal[a] Payback Perio (Years)
1. Properly sized furnace	5	−12	0
2. Bonnet thermostat set 5 °F above room temperature	6	0	0
3. Automatic flue damper	11	5	5
4. Duct insulation increased to 2"	2	1	5
5. Sealed combustion	13	14	8
6. Increase steady state efficiency to 84%	16	9	5
7. Electric Ignition	5	7	13
8. Gas organic fluid absorption heat pump	45	80	16
9. 3, 7	17	13	7
10. 3, 6, 7	24	22	8
11. 2, 5, 6, 7	26	27	9
12. 3, 6	—	—	—
13. 2, 5, 6	—	—	—
14. 1975 model heat pump	—	—	—
15. Compressor indoors (optimized at 28 °F)	—	—	—
16. Two equally sized compressors (optimized at 32 °F)	—	—	—
17. Two speed compressor (optimized at 32 °F)	—	—	—
18. Compressor indoors (optimized at 19 °F)	—	—	—
19. Two equally sized compressors (optimized at 24 °F)	—	—	—
20. Compressor indoors (optimized at 16 °F)	—	—	—
Baseline	84.9 × 10⁶ Btu/yr	$1310	

[a]1975 gas and oil prices assumed. Gas is $1.69/10^5 Btu and oil is $2.80/10^4 Btu.

[b]Based on an end use price of $0.032/kWh ($9.38 10^4 Btu).

[c]Baseline is electirc furnace plus central air conditioning. Located in Cleveland, Ohio.

Source: Dennis O'Neil, *Energy and Cost Analysis of Residential Heating Systems,* Oak Ridge National Laboratory, Oak Ridge, Tenn., July 1978, p. 0.

with reduced firing rate can provide the capability for an estimated minimum improvement in efficiency of approximately 15%.

Variable Firing-Rate Oil Burner. The variable firing-rate burner provides longer burner "on" times for the entire heating season. For moderate heating loads, the burner can operate at a low firing

Oil Systems			Electrical Systems[c]		
% Energy Savings	% Cost Increase	Nominal[a] Payback Period (Years)	% Energy Savings	% Cost Increase (1975 $)	Nominal[b] Payback Period (Years)
4	−13	0	—	—	—
6	0	0	—	—	—
8	5	4	—	—	—
2	1	4	3	1	1
11	8	6	—	—	—
18	8	4	—	—	—
—	—	—	—	—	—
—	—	—	—	—	—
—	—	—	—	—	—
—	—	—	—	—	—
—	—	—	—	—	—
18	12	5	—	—	—
20	16	7	—	—	—
—	—	—	35	11	1
—	—	—	48	13	1
—	—	—	53	16	1
—	—	—	53	21	1
—	—	—	55	27	2
—	—	—	61	31	2
—	—	—	59	62	4
78.5×10^6 Btu/yr	$1815		77.0×10^6 Btu/yr	$2802	

rate (e.g., below 0.5 gal/hr). As the heating load increases the burner firing rate can be increased to satisfy peak demands. The effect of a variable fuel-firing rate is the increase in the total burner "on" time, which reduces the off-cycle losses. However, much developmental work is required to make this retrofit option a viable alternative.

Outside Air Ducting for Combustion Air. The flow of combustion air through the heating unit contributes to the infiltration heating load of the structure, because ambient room air (at 70°F) consumed by the combustion device is replaced by outside air (at a lower temperature). This additional heat load (estimated as 5% of the total

heat load) must be supplied by the heating unit. The use of outside air for combustion eliminates this portion of the heat load
and can significantly reduce the total heat load of the structure.

Balance Air or Water Flow. Adjusting the heating system to provide
enough heat for those areas of the house that need it is also very
important. Dampers, or balancing valves, on central heating systems
should be adjusted to make sure no part of the house is too hot or
too cold. Balancing a home heating system is basically a matter of
lowering the air or water flow, starting with the overheated areas
farthest from the heating plant. If the problem is underheating,
then adjustments should be made to provide more circulation to the
colder rooms.

Temperature Control. Each space or group of rooms having different
demands for cooling or heating should be considered as a separate
zone, and should have an automatic space temperature control. With
the automatic comfort space temperature control, we should be able
to control cooling between 70 and $85^{\circ}F$ and to control heating between
55 and $76^{\circ}F$. Thermostats should be set at $78^{\circ}F$ in the summer and
$65^{\circ}F$ in the winter.

Installing Electric Ignition. In gas furnaces, the pilot light
operates continuously, but the pilot heat is not delivered to the
living area when the circulating fan (blower) is not in operation.
The calculated energy loss for the standing pilot is 4.0×10^{6} Btu
per year. With the installation of electric ignition, this loss is
eliminated.

Advanced Gas Heating Systems. Several types of advanced gas heating systems are being developed for residential applications. These
include the organic fluid absorption heat pump, the Stirling/Rankine
heat pump, and the pulse combustion furnace. The organic fluid
absorption heat pump is expected to be on the market by 1981, and
the Stirling/Rankine heat pump by 1985. No timetable is available
on when the pulse combustion furnace will be marketed. The projected heating SPF for the three advanced systems are given in Table
3-7.

Table 3-7. Estimates of the SPF for
Three Advanced Gas Heating Systems

System	SPF
Organic Fluid Absorption Heat Pump	1.10
Stirling/Rankine Heat Pump	1.41
Pulse Combustion Furnace	0.93

Source: Dennis O'Neil, *Energy and
Cost Analysis of Residential Heating
Systems,* ORNL/CON-25, Oak Ridge Na-
tional Laboratory, Oak Ridge, Tenn.,
July 1978, p. 0.

Heat Pumps. Substantial improvement in energy efficiency can be obtained over electric resistance heat through the use of a heat pump.
The heat pump operates like an air conditioner in reverse, ejecting
warm air into the house and cold air outdoors. Its overall consumption of energy, roughly equal to oil or gas heat, is about one-half
to one-third that of resistance heating. Moreover, the unit can be
used for summer cooling, which saves the extra initial cost of a
separate air conditioner.

Although the greater number of heat pumps are found in the South
and Southwestern United States, the new designs can be used in
northern areas as well. Units should be selected for high effici-
ency--an energy efficiency rate (EER) of at least 10 at standard
test conditions. Where ground water is available, water source heat
pumps should be considered.

Air Conditioning

Approximately 55% of all homes are presently equipped with room air
conditioners or central air conditioning. The performance of air
conditioners now on the market varies greatly, as shown in Fig. 3-2.
The difference in performance reflects both the quality of design
and the cost of the unit. A few units on the market have very high
efficiencies, but this standard of performance is not available in
all size ranges, since manufacturers cannot afford to design conden-
sers that are optimally suited for all compressors to which they may
be attached.

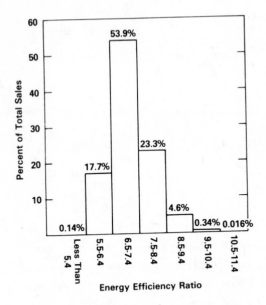

Figure 3-2. *Air conditioning EER of heat pumps and central air con-*
ditioning units shipped in 1977 (includes only units of under 135,000
Btu-hr). (From Air Conditioning and Refrigeration Institute.)

Selecting high-performance equipment is the best way to reduce
energy consumption in air conditioning equipment. The difference in
the efficiencies of air conditioners can be compared according to
their EERs. The EER is the capacity measured in Btu's per hour,
divided by the energy used, measured in watts.

$$EER = \frac{Btu/hr}{watts}$$

The higher the EER number, the more efficient the air conditioner.
For room air conditioning units the EER ranges from about 4 to 12.
For central air conditioners, the EER ranges from about 5 to 11.
Comparing two 36,000-Btu units with EERs of 5 and 8, respectively,
we find that the energy used by each is as shown below:

$$Watts = \frac{Btu/hr}{EER} = \frac{36,000}{5} = 7200 \ W$$

for the standard performance unit and

$$\text{Watts} = \frac{36,000}{8} = 4500 \text{ W}$$

for the high-performance unit. This means that the additional energy required by the unit with an EER of 5 is

$$\frac{7200 \text{ W}}{4500 \text{ W}} = 1.60$$

or 60% greater than that of the high-performance unit.

The performance of electric cooling and heat pump systems also varies as functions of the temperatures and humidities of both the inside and outside air. This is because the theoretical capacity of a unit varies as a function of these parameters, and because most small units must be either fully on or fully off. The load control achieved by "cycling" the system from full capacity to zero output requires heating or cooling large parts of the system before useful space conditioning can be performed. Using energy to heat or cool the units decreases the system's efficiency.

A number of straightforward changes can be made to improve the performance of air conditioners without changing the basic design. Some of the steps being taken to improve performance include the incorporation of:

1. More efficient compressors
2. Two compressors or multiple-speed compressors to improve part-load efficiency
3. Improved heat exchangers for both the condensor and evaporator
4. More efficient motors to drive the compressor and fan
5. Automatic cycling of the fan with the compressor
6. Improved airflow

Table 3-8 shows how these room air conditioner design options can save a significant amount of energy.

Looking further into the future, a number of systems that could increase the coefficient of performance (COP) of air conditioning systems and heat pumps by as much as 50% have been proposed. The speed with which these new units appear on the market will depend

Table 3-8. Room Air Conditioner Design Options to Improve Efficiency

Design Option	Production-Weighted Energy Savings (%)
Install a Switch to Cycle Fan with Compressor	5
Improve Fan Motor Efficiency	3
Improve Compressor Efficiency	5
Improve Cycle Efficiency	10
Total	23[a]

[a]The production-weighted energy savings of 23% corresponds to an efficiency improvement of 30%.

Source: G. E. Liepins, M. A. Smith, A. B. Rose, and K. Haygood, *Buildings Energy Use Data Book,* 1st ed., ORNL-5356, Oak Ridge National Laboratory, Oak Ridge, Tenn., April 1978, p. 83.

strongly on the manufacturer's perception of whether the public is willing to invest in equipment that can reduce its annual operating expenses on a long-term basis.

HOT WATER HEATERS

Hot water heaters account for about 15% of residential and 3% of the total national energy consumption. American homes consume an average of 26,000 gal of hot water each year, and the amount of energy it takes to heat this water is more than that consumed by the refrigerator, freezer, clothes dryer, television, kitchen oven and range, lights, and central air conditioning combined. In fact, the amount of energy needed to heat water is second only to the amount used to heat the house.

Two types of water heaters are sold in the United States: electric and gas-fired. In both types, an insulated tank of water is continually maintained at the desired delivery temperature (generally about 140°F). For electric water heaters, a 52-gal tank which uses two 4.5-kW immersion electric heaters is typical. Forty gallon storage tanks are typical for the gas-fired units, which generally have main burner units of 45,000 Btu/hr size and pilot lights of 700 Btu/hr size.

The major factors that affect energy consumption in water heaters are hot water usage, thermostat setting, temperature of the make-up water, and ambient air temperature. Obviously, the most important of these is the amount of hot water actually used.

The largest heat loss in an electric water heater is conduction through jacket walls (14% of energy input). An additional 5% is lost through the distribution pipe of 7.6 m (25 ft) length. The remaining 81% of energy input is used to heat water. In a gas water heater, conduction losses through jacket and distribution pipe are 12% and 3%, respectively; and 33% is lost up the flue (as a result of main burner and pilot light operation). Only 52% of energy input to a gas water heater is used to heat water (2, p. iii) (see Figs. 3-3 and 3-4).

Operation and Maintenance Tips

Energy usage for hot water heating can be reduced at low or no cost in many ways. The major energy saver is to reduce the amount of hot water used. Other steps that can be taken are discussed as follows.

Reduce Temperature Setting. Most water heaters have adjustable thermostats, but most people keep them at the factory setting, usually 140 or 150°F. By lowering the thermostat setting 10°, energy consumption can be decreased by about 5%. Consider experimenting: Lower the thermostat by 20°, and if the automatic dishwasher and clothes washer perform adequately at that temperature, hot water energy consumption can be reduced by 10% (for electric) or 9% (for gas).

Reduce Hot Water for Showers. The amount of hot water used for showering depends on three factors: the temperature of the water, the duration of the shower, and the rate of water flow. Adjusting any one of these factors can result in considerable energy savings, since bathing generally accounts for a substantial portion of hot water use.

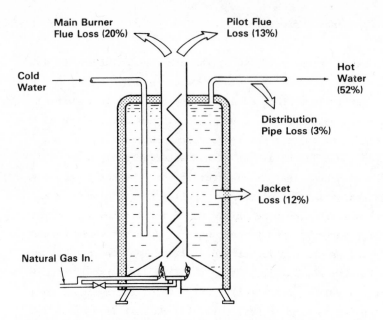

Figure 3-3. Energy flows in a 40-gal water heater. (From Air Force Energy Conservation Handbook, vol. 1, National Bureau of Standards, January 1977, p. 10.)

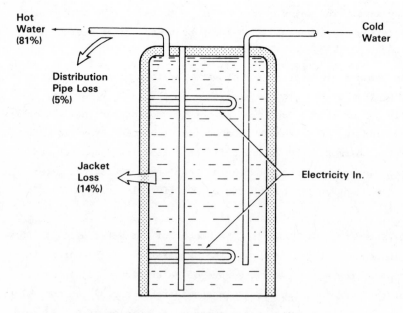

Figure 3-4. Energy flows in a 50-gal electric water heater. (From Air Force Energy Conservation Handbook, vol. 1, National Bureau of Standards, January 1977, p. 10.)

Use Automatic Dishwasher Wisely. Running the automatic dishwasher
only when full will reduce the frequency of use, which saves both
hot water and electric energy.

Use Clothes Washer Wisely. Two basic ways of reducing the amount of
hot water consumed by automatic clothes washers are washing full
loads only and using less hot water by lowering temperature settings.
The savings can be substantial. Studies show that eight washloads
per week using a hot wash/warm rinse setting consume an average
13,104 gal of hot water each year--about half of the average household's total annual hot water consumption. By changing the setting
to hot wash/cold rinse, annual hot water consumption falls to approximately 8650 gal, or a savings of some 4450 gal per year (3).
In fact, there is no reason for not using a cold rinse, since the
temperature of the rinse water has no effect on the cleaning ability of the washing machine. By using a cold wash/cold rinse setting, hot water usage falls to zero, which means that the homeowner
may be able to cut his total household hot water consumption in half.

Insulate Storage Tanks and Pipes. By adding $1\frac{1}{2}$ in of insulation to
an electric water heater, a saving of about 450 kWh/year can be
achieved (4). This means a savings of $18 at 4¢/kWh. For gas-fired
systems, the savings amounts to 3600 ft^3 of gas, and at $0.231/therm
(100 ft^3), it amounts to about $8 per year. Hot water piping should
also be insulated. The cost-effectiveness of piping insulation depends on several factors, such as water and ambient air temperature,
accessibility of piping and, in particular, water use habits. It
is particularly important to insulate vertical runs of piping adjacent to the water heater.

Turn Off During Vacation. If one is leaving home for more than just
a few days, one can save energy by turning off the water heater completely, or at least by turning the thermostat to the lowest setting.

Energy Conserving Design Changes

In an ideal hot water heater, the delivery per day of 71.4 gal of
hot water, heated to 85°F, would require 55,000 Btu/day (51,000 Btu
to heat the water at 100% efficiency and 4000 Btu minimum standby

loss). Compared to this ideal, a 52-gal electric water heater uses 61,000 Btu/day of electric energy, and a 40-gal gas-fired water heater uses 102,000 Btu/day of natural gas energy. The difference between the ideal requirement of 55,000 Btu and the real current usages of 61,400 and 102,000 Btu for electric and gas water heaters, respectively, represents the maximum potential for energy savings through design changes of existing models (5, p. xxi).

A number of energy-conserving designs for improved efficiencies of both electric and gas water heaters are available (2, pp. 16-23). These options include increasing jacket insulation thickness and reducing jacket insulation thermal conductivity. For the gas water heater, whose efficiency (52%) is lower than that of an electric water heater (81%) because of flue losses, other design changes include reducing the pilot rate or eliminating the pilot, adding an electric ignitor and flue closure, and increasing flue baffling.

Increase Jacket Insulation Thickness. Electric water heaters generally have 2 in. of fiberglass jacket insulation, and gas water heaters generally have only 1 in. of insulation. Since one of the largest heat losses in both electric and gas water heaters is through jacket walls, an increase in jacket insulation thickness will reduce energy loss significantly. Table 3-9 shows the energy savings, added cost, and pay-back period for increased insulation of electric

Table 3-9. Energy and Economic Effects of Increased Jacket Insulation Thickness*

Insulation Thickness (Inch)	Electric			Gas		
	% Energy Savings	Added Cost (1975-$)	Payback Period[a] (Yr)	% Energy Savings	Added Cost (1975-$)	Payback Period[a] (Yr)
2	—	—	—	7	4.9	1.1
3	4	6.1	0.7	10	10.9	1.9
4	7	12.1	1.0	10	17.0	2.7
5	7	19.4	1.4	11	24.2	3.6

*Based on 1975 fuel prices.
Source: Ref. 6, p. 18.

and gas water heaters. An increase in insulation thickness beyond 4 in yields only slight energy savings.

Improve Jacket Insulation Thermal Conductivity. Foamed-in place urethane provides better insulation than fiberglass because of its lower thermal conductivity. Table 3-10 shows energy savings, added cost, and pay-back period for use of urethane foam as jacket insulation.

Another alternative for reducing thermal conductivity is the installation of "double-density" fiberglass. This provides an improvement in energy savings over single-density fiberglass but is not as effective as urethane foam.

Reduce Pilot Rate. The pilot light in a gas water heater, which operates continuously, serves as both a heat source to tank water and an ignitor for the main burner. However, when the main burner is off, only 22% of the pilot energy heats water; the remainder escapes up the flue. The energy lost by the operation of the pilot is decreased if the pilot rate is reduced. This can be accomplished by reducing the pilot orifice area and can result in an energy savings of about 6%.

Eliminate Pilot-Add Electric Ignitor and Flue Closure. For a gas water heater, flue loss during standby accounts for 13% of the energy input. Using an electric ignitor in place of the pilot, without

Table 3-10. Energy and Economic Effects of Reduced Jacket Insulation Thermal Conductivity

Insulation Thickness (Inch)	Electric			Gas		
	% Energy Savings	Added Cost (1975-$)	Payback Period (Yr)	% Energy Savings	Added Cost (1975-$)	Payback Period (Yr)
1	—	—	—	9	6.8	1.2
2	8	9.5	0.5	12	13.0	1.8
3	10	17.5	0.8	13	17.5	2.3
4	11%	22.0	1.0	13	22.0	2.8
5	11%	26.5	1.1	13	26.5	3.3

Source: Ref. 6, p. 18.

using the damper, can eliminate flue loss of heat generated by the pilot. However, the hot water in the tank will continue to lose heat to the cooler gases in the flue, which then flow up the stack. Thus, the energy saved by merely replacing the pilot with an electric ignitor is negligible.

When an electric ignitor and automatic damper are installed together, the flue loss is eliminated because the damper prevents hot air from escaping during standby. Energy savings gained by using the electric ignitor and flue damper are about 14%, but the pay-back period is so long that this is not a cost-effective design change.

Reduce Excess Air for Combustion (Increase Flue Baffling). The loss of heat up the flue of a gas water heater is proportional to the amount of excess air used for combustion, because this air is heated during combustion and then escapes up the stack. Restricting air flow in the flue reduces the amount of excess air and lowers the flue gas exit temperature. This can be achieved by increasing flue baffling or by repositioning the baffling in a manner that hinders airflow. However, the amount of baffling that can be added to the flue is limited, because too much airflow restriction would eliminate the pressure difference (necessary for natural draft) between entering air and exiting gas, and combustion gases would leak into the surrounding air. Flue baffling can result in approximately 6% energy reduction.

Heat Pump Water Heaters (6). The recent advances in heat pump technology has generated an interest in applying heat pumps to domestic water heating. The general principle underlying the coupling of heat pumps to water heaters is similar to that of heat pumps used in space heating. Heat is extracted from cool ambient air and "pumped" into a medium at a higher temperature. For a heat pump used in space heating, this medium is the conditioned air inside a structure. For a heat pump water heater, this medium is water.

Two heat pump designs which call for small heat pump units installed together with a conventional hot water tank are feasible. One uses a standard Rankine cycle, and the other a Brayton cycle.

The Rankine cycle heat pump water heater uses a motor compressor,
an evaporator, an expansion valve, refrigerant, and a control system similar to those used by space heating heat pumps. The heat pump
cabinet is built as part of the water heater, and the heat pump water
heater fits into the same floor space a conventional water heater
does. Preliminary results show that the Rankine cycle heat pump
has a SPF of at least 2 and, under certain conditions, as high as 3.
The performance of this heat pump water heater varies with the source
of heat (ambient air), water supply temperature, and hot water temperature. The performance improves as the source air temperature
increases; it decreases as the water supply or hot water temperature
increases.

The Brayton cycle heat pump consists of a reciprocating compres-
sor/expander, an air-to-water heat exchanger, an electric motor, a
water circulation pump, and controls. The system fits into a cyclin-
drical package under the water tank. The estimated SPF and initial
cost of this heat pump is expected to be 1.7 and $430, respectively.
Again, performance improves as the source air temperature increases,
and decreases as the water supply or hot water temperature increases.

Table 3-11 shows the expected pay-back periods for the two heat
pump water heaters, in comparison to a conventional electric water
heater, in three cities. Because of the short pay-back period, heat
pumps should provide an attractive alternative to conventional electric water heaters.

The heat pump water heater achieves its greatest feasibility in
areas where the cost of electricity is high and in homes where hot

*Table 3-11. Expected Energy Savings and Economics for Heat Pump
Water Heaters*

Heat Pump	% Energy Savings	% Cost Increase	Payback Period (Years)		
			Portland	Kansas City	New York
Rankine Cycle	50	157	3.4	2.1	1.0
Brayton Cycle	41	173	4.5	2.7	1.4

Source: Ref. 6, p. 9.

water consumption is also high. The southern portion of the country
and other areas such as the Pacific Northwest and Hawaii are good
locations because of year-round moderate climates. The New England
area can achieve large energy benefits because of the large number
of homes with electric water heaters and the higher efficiency of
heat pump water heaters (see Fig. 3-5).

In the South, installing a heat pump water heater which could
also provide the fringe benefits of air conditioning and dehumidifi-
cation for most of the year can represent a savings of one-third
the installation cost of the device, plus two-thirds the cost of
water heating. The same principle would be true in the summer in
the North. However, in the winter, the heat pump would take heat
from the house, which must then be replaced by the home heating system. This negative factor is compounded if the unit were to be installed in a conditioned space, such as a heated basement or utility
room.

Summary

If the energy-saving options described previously are taken advantage of, electric water heaters can approach the ideal energy consumption level of 55,000 Btu/day (for the specified 71.4-gal, 85°F-
rise hot water usage). This represents a 10% savings over the 1972

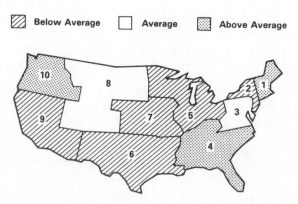

Figure 3-5. Heat pump energy savings. (From Ref. 6, p. 30.)

reference point of 61,400 Btu/day (18.0 kWh/day). Gas-fired water
heaters can achieve 85,000 Btu/day (a 17% reduction) by simple changes
in the amount of insulation used, the thermostat set point, the
amounts of baffling used, and by reducing the pilot rate. It is
feasible for gas-fired water heaters to reach 66,000 Btu/day (35%
savings) by adding forced draft, flue dampers, and electric ignition. In addition, significantly greater savings could accrue if
simple steps are taken to reduce hot water usage in the home.

REFRIGERATORS AND FREEZERS

Virtually every American household has a refrigerator of some kind,
and almost half of them have separate freezers as well. Saturation
may actually exceed 100%, because many families own two refirgera-
tors (e.g., for use in a basement or in a second home). If incomes
continue to rise, refrigerator ownership will probably continue to
increase faster than household growth.

Refrigerators are classified as either single-door, top-freezer,
bottom-freezer, or side-by-side types. Top-freezer models are the
most popular because they take up less floorspace per unit volume
and cost less than side-by-side units. In 1975, about 80% of the
refrigerators sold were top-freezers and 20% were side-by-side (5).
Refrigerators have one of three types of defrost systems--manual,
partial, or automatic. Most new refrigerators have automatic defrost. In 1965, 48% of new refrigerators had automatic defrost;
this percentage rose steadily to 73% in 1975.

Approximately 58% of the life-cycle costs of owning and operating a refrigerator are due to electricity consumption. The relative importance of operating costs is due to both the long average
lifetime of refrigerators (about 15 years) and the high electricity
consumption. It is estimated that refrigerator-freezers account
for almost 7% of all energy consumed in the residential sector of
the American economy.

Refrigerator energy use depends primarily on the design of the
unit: size, defrost option, and door configuration. Energy use is

also influenced by such operational characteristics as room temperature, food load, and the number of door openings. Because of these design and operational factors, estimates of refrigerator electricity use vary considerably.

Refrigeration equipment consumes energy to achieve specific temperature settings by removing heat from the storage compartment. Figure 3-6 shows where the heat comes from. Energy is also consumed by heaters and fans used for both automatic defrosting and reducing frost and moisture buildup. The schematic diagram in Fig. 3-7 shows how energy is used in refrigerators and indicates the percentage of electrical energy consumption for which each compartment is responsible.

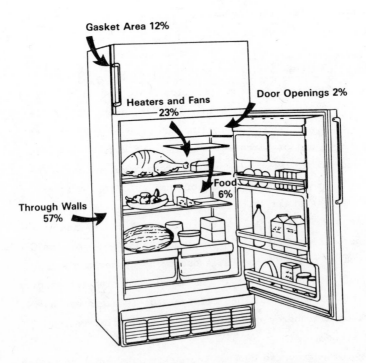

Figure 3-6. Refrigerator heat gains. (From Ref. 5, p. 11.)

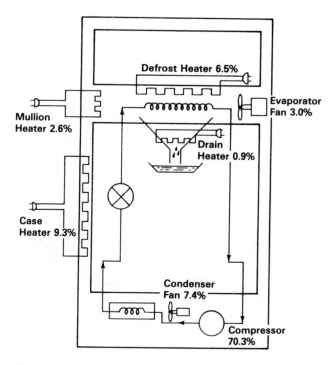

Figure 3-7. Refrigerator electricity consumption. (From Ref. 5, p. 12.)

Operating and Maintenance Practices

Good operating and maintenance practices provide large opportunities for energy conservation. Some of the major practices are suggested in the sections that follows.

Use the Most Efficient Temperature Setting. The ideal temperature setting provides just the amount of refrigeration needed. But what "just right" is depends on many different factors, such as the amount and type of food stored and the frequency with which the doors are opened. In most cases, refrigerator temperatures should be kept at from 35 to 40°F; freezer temperatures from 0 to 8°F. Using settings lower than those actually needed wastes energy. For every degree that the temperature setting is lowered, energy consumption increases by 5% (5, p. 16).

Check Door Seals. Door seals prevent the loss of cold air when the refrigerator or freezer is closed. But seals cannot work efficiently if they are dirty, loose, broken, or worn. Therefore, door seals should be inspected frequently and replaced or repaired as necessary.

Clean Condenser Coils. Condenser coils are located either behind the bottom grill panel or in back of the refrigerator. The dust and dirt that collects on them acts as undesirable insulation that reduces efficiency and wastes energy. For top performance, coils should be cleaned at least twice a year by using the nozzle attachment of a vacuum cleaner.

Defrost Regularly. If the refrigerator-freezer has manual or partial defrost systems, it should be defrosted whenever frost build-up exceeds ¼ in. Like dirt, frost also acts as undesirable insulation, necessitating more energy consumption to maintain a given temperature.

Keep Refrigerators and Freezers Full. Keeping refrigerators and freezers full means that there is less air inside and that less cold air is lost when the door is opened.

Reduce Door Openings. Every time a refrigerator or freezer door is opened, warm air in the room enters the storage space. This raises the temperature inside and activates the refrigeration equipment, consuming energy.

Energy-Conserving Design Changes

The following design changes can improve a refrigerator's energy efficiency: changing insulation type and/or thickness, removing the evaporator fan motor from the refrigerated space, using an anti-sweat heater switch, eliminating the frost-free feature, improving

compressor efficiency, and increasing the condenser and/or evaporator heat transfer surface areas.

In a study for the Federal Energy Administration, Arthur D.
Little, Inc. examined 30 energy-saving design options for two
specific models (A and B) of the typical 16-ft^3 refrigerator-freezer
(10, p. 16). Model A had fiberglass insulation, and model B was a
thin-walled foam refrigerator. Twenty options with energy savings
sufficient to offset added first costs over the lifetime of the unit
were considered. These are listed in Table 3-12. The Little Company estimated that fourteen of these options can be implemented by
1980, three by 1982, and the remaining three at some undetermined
time in the future.

All of the options were evaluated with reference to a standard-
duty cycle. The added first cost is the production-volume-weighted
average of the added cost, based on three sizes of manufacturing
facilities. The life-cycle cost is a straight-time, zero-discounted,
present-worth savings for the 10-year useful life of a refrigerator
and is based on a $.04/kWh energy cost.

Energy savings provided by combining all options which can be
implemented for representative refrigerator models by 1980 are given
in Table 3-13.

Summary

Good operating and maintenance practices as well as new designs can
result in substantial energy savings in refrigerators and refrigerator-freezers. Recent studies have indicated that savings of 40 to
50% can be achieved. The price increases for the improvements would
be minimal, and the life-cycle cost to the consumer would be significantly less. Figure 3-8 provides an indication of energy use
versus retail price for various design changes for a 16-ft^3 top-
freezer refrigerator. The potential for energy conservation is substantial.

Table 3-12. Evaluation of Energy Saving Options for 16-ft³ Automatic Defrost, Combination Refrigerator-Freezer

Time Frame In Which Options Are Implementable	Description
1980 or Before	**Insulation System**
	Replace 2.5-inch Fiberglass with 2.5-inch Foam[+]
	Increase 1.75-inch Foam to 3-inch Foam[++]
	Improved Door Closure Area
	Better Insulate Interchanger
	Substitute Natural Convection Condenser
	Compressor
	Motor Modifications
	Reduced Suction Gas Heating
	High-Side Type
	Low-Side Type
	Evaporator
	Increased NTU
	Place Evaporator Fan Outside Cold Space
	Improve Evaporator Fan/Motor
	Shut Evaporator Fan Off During Door Openings
	Defrost and Moisture Control
	Hot Gas Defrost
	Reduce Moisture Heaters
	(In Combination with Insulation Changes)
1980-1982	**Compressor**
	Improved Valves/Ports
	Improved Mechanical Efficiency
	Reciprocating
	Rotary
Beyond 1982	Evacuated Power Insulation
	Expander:
	Instead of Capillary as Motor for Evaporator Fan
	Sequential Control

*Model A presently has fiberglass insulation.

**Model B has thin-walled foam.

Source: Ref. 5, p. xvi.

Estimated Energy Savings (%)	Added First Cost (%)	Life-Cycle Cost Saving
20	$13.54	$128.06
20	$12.48	$125.12
1	$ 2.00	$ 4.70
1	$ $1.61	$ 45.48
	$ 5.00	Done Only in Combination with Above
12	$20.12	$ 58.73
2	$ 1.00	$ 10.60
6	$ 3.00	$ 39.70
5	$11.33	$ 21.97
3	$ 0.90	$ 18.10
4	$ 2.40	$ 39.31
3	$ 0.13	$ 19.65
6	$26.45	$ 14.05
5	$ 0	$ 31.80
6	$ 0.73	$ 38.97
8	$20.30	$ 47.90
12	Specific Hardward Realization Unknown; No Cost Estimate Made	
40		
9	Specific Hardware Realization Unknown; No Cost Estimates Made	
15		

Table 3-13. Energy Savings Achievable for Representative Refrigerator Models

Model	Baseline Energy Use (kwh/day)	% Energy Savings With All Options Implementable by 1980
16 Cubic Foot, Model A	4.57	44
16 Cubic Foot, Model B	4.51	43
12 Cubic Foot, Partial Automatic Defrost Combination	2.16	43
22 Cubic Foot, Side-by-side Automatic Defrost	5.16	39
22 Cubic Foot Chest Freezer, Manual Defrost	3.04	56
17 Cubic Foot Upright Freezer, Automatic Defrost	5.53	53

Source: Ref. 5, p. xvii.

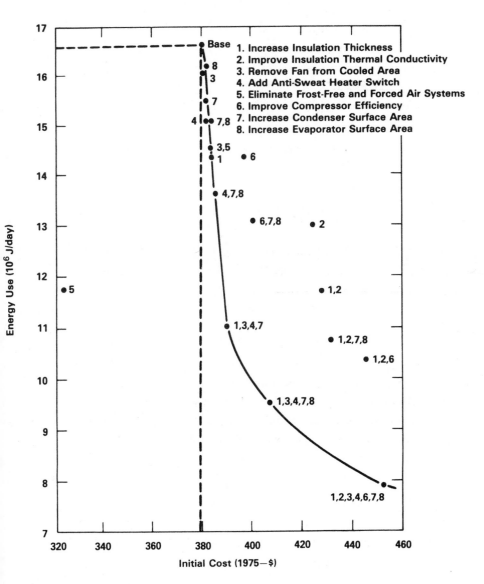

Figure 3-8. Energy use versus retail price for a top freezer-refrigerator. (From Ref. 2, p. 5.)

RANGES AND OVENS

Virtually all American households have ranges and ovens. These
ranges and ovens account for about 5% of the residential and 1% of
the total national energy consumption. As with other appliances,
proper use and maintenance can have significant impact on the amount
of energy consumed.

Some measures which can reduce cooking energy consumption include:

1. Using the right equipment for the job (e.g., not toasting one
   slice of bread in the oven).
2. Using the oven and broiler to full capacity. The same amount
   of energy is consumed, whether all or just part of the space is
   being used.
3. Thawing frozen food before cooking it. This saves energy, because less heat will be needed to cook it. Frozen food thawed
   in the refrigerator will reduce the energy consumed by the refrigerator to keep things cold.
4. Turning off equipment which is not needed.
5. Keeping appliances clean. Dirty appliances consume more energy
   because dirt and grease build up on heat transfer surfaces.
6. Using a pressure cooker whenever possible.
7. Using microwave cooking whenever possible. Although the microwave oven saves energy when it replaces surface cooking, it
   saves the greatest amount of energy when it replaces oven cooking.

Table 3-14 shows the annual koliwatt-hours used to prepare 27
foods from a menu suitable for microwave cooking. Foods were cooked
on three appliances and compared. As can be readily seen, the
energy efficiency of the microwave oven was 50% better than that of
the 1976 electric range and 69% better than that of the 1965 vintage
electric range.

Table 3-14. Percentage of Savings From Microwave
Versus Conventional Electric Range

	1976 Electric	Microwave	% Savings
From Cooktop	211 kwh	166 kwh	21
From Oven	248 kwh	62 kwh	75
TOTAL:	459	228	50
	1965 Electric	Microwave	% Savings
From Cooktop	334 kwh	166 kwh	50
From Oven	393 kwh	62 kwh	84
TOTAL:	724	228	69

Source: U.S. Department of Energy, Proceeding of the
Conference on Major Home Appliance Technology for
Energy Conservation, Feb. 27 to Mar. 1, 1978, Purdue
University, June 1978, p. 45.

Energy-Conserving Design Changes

A number of improvements can be made to enhance the energy efficiency of ovens and ranges. Table 3-15 presents a review of the available literature on potential energy efficiency improvements.

Eliminating Standing Pilots. Gas ranges equipped with automatic
ignition devices consume 25 to 47% less energy than those with standing pilot lights. With the retail cost of electric ignition systems
estimated to be between $25 and $50, the trend in the appliance
industry seems to be toward more use of automatic ignition, especially on the more expensive models of ranges.

Forced Convection Ovens. A limited number of forced convection
ovens, both gas and electric, are now available. Their potential
efficiency improvements compared to regular ovens are greater for
gas ovens than they are for electric ovens (between 8 and 20% for
gas and 1 and 5% for electric). The estimated retail cost for this

Table 3-15. Summary of Technological Improvements in Ovens and Ranges

		Booz, Allen & Hamilton	Unified Industries	Frigidaire	Tappan
Eliminate Standing Pilot Light	Efficiency Improvement	31%	13.9%		25%
	Cost Impacts	$15-$25 Mfrs. Cost	$30 Retail Cost		$50 Retail Cost
Forced Air Convection	Efficiency Improvement	11.3%-16.5%	18% Gas		8% Gas
	Cost Impacts	$100-$125	$75-$100		$75-$100
Improve Oven Insulation	Efficiency Improvement	10-20% Elec. 1% Gas	5.93% (Incl. Door Seals) Elec.	2% Elec.	5%
	Cost Impacts	$9-$15 Mfrs. Cost	$7 .10 Retail Cost		
Improve Oven Door Seals	Efficiency Improvement	2% Elec. 0.2% Gas		5%	
	Cost Impacts				
Reduce Wattage/ BTUs on Surface Units	Efficiency Improvement	5% Elec. 1.5% Gas		7% Elec.	
	Cost Impacts	0			
Reflector Pans	Energy Savings or Efficiency Improvement	3.5% Elec.			
	Cost Impacts	$4-5 Retail Cost			

Source: Enviro-Management and Research, Promotional Information for Energy Conservation, U.S. Department of Energy, March 1978, pp. 3-23.

Roper	General Electric	Magic Chef	Whirlpool	Comments
20-30%		25%		Increased Acceptance
$44.50 Retail Cost		$30-$35 Retail Cost		Potential Additional Maintenance Cost
0% Elec. 8% Gas		0-5% Elec.	1.2% Elec.	Currently Available
$75-$100				Economically Infeasible Some Features May Be Patented
5% Elec. 1.2% Gas	2-3% Elec.		5.5% Elec.	May Not Apply to Self Clean Ovens
	.75%			
	0%		4%	May Require Redesign of Production Processes
	0-2% For Those With Porcelain			Already Widely Used

feature ranges from $75 to $125, thus making its economic desirability uncertain, especially for electric ovens.

Improving Oven Insulation. Increased insulation would improve the
energy efficiency of electric ovens (between 2 and 10%) more than
it would gas ovens (between 0 and 1%). The estimated retail cost of
this improvement ($9 to $15), which applies only to non-self-cleaning
ovens, indicates a high probability of economic desirability from
the consumer's point of view.

Improving Door Seals. Improving door seals, like increasing insulation, can save between 2 and 5% in the energy consumption of electric ranges. The savings is less than 1% for gas ranges that do not
already have door seals.

*Reducing Wattage on Electric Surface Units/Reducing Btus for Gas
Burners.* Reducing the wattage on electric burners can save up to 7%
in the energy consumption of electric ranges, and reducing the Btus
of gas burners can save up to 1.5% for gas units. Implementation of
this change for electric units might entail a change in production
processes, thereby increasing the price, whereas for gas units,
there should be no increase in cost.

Using Reflector Pans. Most studies agree that the use of reflector
pans increase the energy efficiency of surface units. For those
units with porcelain pans, a 2% improvement could be achieved by
utilizing reflector pans. The cost of implementing this feature is
about $5.

Summary

On the basis of past studies, energy efficiency improvements of 40
to 50% for gas ranges and 10 to 20% for electric ranges seem feasible,
if the technologically and economically justified changes are implemented. However, industry sources indicate that these estimates are
high and that only half of these savings can actually be achieved.

CLOTHES WASHERS AND DRYERS

Clothes washers and dryers account for about 2% of the energy consumed in an average residence each year. This figure does not include the energy needed to heat the water used by washing machines.
Although some washers and dryers are more efficient than others, the
amount of energy they consume is primarily determined by how they
are operated.

The amount of energy used for clothes washing is, to a large
extent, dependent on the amount of hot water used and on the temperature and fill control settings selected. Temperature controls
affect both the wash and rinse cycles; fill controls indicate the
load size that can be cleaned most effectively. Most consumers do
not use maximum capacities in their washers and dryers. According
to one study, the average load put into a regular-size (14 lb)
washer is 5.4 lb, and the average load put into a large capacity
(16 to 20 lb) model is only 5.9 lb (7). Despite the tendency to
underload, studies show that consumers use full water fills in
regular-size machines 80% of the time, and in large-capacity machines 60% of the time.

The size of a clothes dryer is determined by the size of its
drum. There are no standard sizes: the general categories available are compact, regular, and large. The size of the unit has very
little apparent effect on the efficiency of its energy use.

By and large, the energy consumed for washing and drying clothes
depends not so much on the appliance as it does on how they are used.
As a general rule, knowing the proper use of all of the appliance's
controls and taking advantage of its other features will save energy
and money. Energy efficiency can be achieved in other ways:

1. Whenever possible, load the washer or dryer to the maximum cap-
    acity that can be handled without sacrificing satisfactory re-
    sults. This saves energy, since in most cases almost as much
    energy is consumed in handling a partial load as is consumed in
    handling a full load.

2. Wash and dry during off-peak periods. In most parts of the
    country, electrical demands are highest during weekday hours
    and lowest during evenings and weekends. Washing and drying
    during the off-peak periods enables the electrical utility com-
    pany to provide electricity most efficiently.
3. Use cold water rinses. There is no reason whatsoever for not
    using cold rinses, since the temperature of the rinse water has
    no effect on the cleaning ability of the washing machine. Using
    a hot or warm rinse only wastes energy.
4. Avoid hot water washing, if possible.
5. Install an electric ignitor, if the gas dryer has a pilot light.
    In most cases, this change will be cost-effective, since it can
    reduce the average dryer energy consumption by some 40% (8).
6. Install a suds saver to the washer, if possible. A suds saver
    is a device that stores the water used from one wash for im-
    mediate use in a second wash load. Consequently, less hot water
    is needed to bring the water up to the desired temperature.

INTEGRATED APPLIANCES

Of the several sources of "waste heat" in a home, some could be effectively utilized by means of integrated appliances, appliances
that can recover heat (that is now completely wasted) for either
space heating or water heating (e.g., air conditioner/hot water
heaters and refrigerator/water heaters). A study by Arthur D. Little,
Inc. discusses a number of promising systems (9), four of which look
especially promising. These include an air conditioner/water heater
a furnace/water heater, a refrigerator/water heater, and a drain
water heater recovery system. Only the air conditioner/water heater
is commercially available at this time.

All of these systems must compete with the improvements being
made in space and water heating, since it is impractical that, in a
single house, more than one system for augmenting hot water production be implemented. Moreover, the marketing of integrated appliances will be constrained by the cost of heat recovery and the demand for the heating of hot water.

APPLIANCE EFFICIENCY STANDARDS

Congress established appliance efficiency standards to encourage manufacturers to produce and consumers to purchase more energy-efficient appliances than those being produced and purchased today. In order to meet these objectives, two interrelated strategies were implemented.

The first strategy, a part of the Energy Policy and Conservation Act (EPCA), called for the promulgation of voluntary efficiency improvement targets, which were to be achieved by 1980. A reporting and monitoring system was established by the Department of Energy (DOE). In April 1977, the President proposed the National Energy Plan, which replaced the voluntary target program with a mandatory minimum efficiency standards program. This was enacted into law by the National Energy Conservation Policy Act of 1978. The Act prescribes energy efficiency standards to be achieved by December 1980 for the following appliances: refrigerators and refrigerator-freezers, freezers, clothes dryers, water heaters, room air conditioners, home heating equipment, kitchen ranges and ovens, central air conditioners, and furnaces. In addition, standards for dishwashers, televisions, clothes washers, humidifiers, and dehumidifiers are to be established by November 1981.

The second strategy, aimed at assiting consumers to make better purchasing decisions, involved the development of a labeling program which required manufacturers to label each consumer product with energy consumption information. In conjunction with the issuance of the labeling rules, the DOE developed a consumer education program to enhance customer awareness of the labels and to give the consumer a better understanding of the information provided on the labels: If better informed, the consumer would theoretically be more inclined to engage in comparison shopping and to demand more efficient products.

To support these strategies the DOE developed test procedures to determine energy efficiency and estimated annual operating costs for each consumer product to assist consumers in making purchasing decisions. The test procedures are also used as the basis for energy

Table 3-16. Likely Classes and Tentative Determinations of Maximum Technologically Feasible Energy Efficiency Levels*

Covered Product Type	Class	Preliminary Maximum Technologically Feasible Energy Efficiency Level
Refrigerators and Refrigerator-Freezers	Electric, Manually Defrosted, 15' Freezer	10.4 Ft.³/kWh-Day (EF)
	Electric, Manually Defrosted, 5' Freezer	10.1 Ft.³/kWh-Day (EF)
	Electric, Automatic Defrost	6.6 Ft.³/kWh-Day (EF)
Freezers	Manual Defrost, Chest	16.9 Ft.³/kWh-Day (EF)
	Manual Defrost, Upright	13.9 Ft.³/kWh-Day (EF)
	Automatic Defrost	9.1 Ft.³/kWh-Day (EF)
Clothes Dryers	Electric, Standard	2.77 Lb./dWh (EF)
	Electric, Compact	2.61 Lb./dWh (EF)
	Gas	2.46 Lb./dWh (EF)
Water Heaters	Electric	0.89 (EF)
	Gas	0.59 (EF)
	Oil	0.50 (EF)
Room Air Conditioners	Window and through the Wall (with Outdoor Side Louvers)	11.6 Btu/Watt-Hour (EER)
	Through the Wall (No Outdoor Side Louvers)	7.5 Btu/Watt-Hour (EER)
	Packaged Terminal	8.7 Btu/Watt-Hour (EER)
	Reverse Cycle	8.8 Btu/Watt-Hour (EER)
Home Heating Equipment, Not Including Furnaces	Electric, Primary and Supplementary	100% Efficiency
	Gas, Gravity, Vented Room Heater	58% (AFUE)
	Gas, Forced Air, Vented Room Heater	74% (AFUE)
	Gas, Gravity, Vented Wall Furnace	60% (AFUE)
	Gas, Forced Air, Vented Wall Furnace	70% (AFUE)
	Gas, Gravity, Vented Floor Furnace	70% (AFUE)
	Gas, Forced Air, Vented Floor Furnace	(AFUE)
	Oil, Gravity, Vented Room Heater	(AFUE)
	Oil, Forced Air, Vented Room Heater	(AFUE)
	Oil, Gravity, Vented Wall Furnace	(AFUE)
	Oil, Forced Air, Vented Wall Furnace	(AFUE)
	Oil, Gravity, Vented Floor Furnace	(AFUE)
	Oil, Forced Air, Vented Floor Furnace	(AFUE)

Kitchen Ranges and Ovens

Microwave Oven	44% (EF)
Electric Cooking Top	79% (EF)
Electric Oven	16% (EF)
Electric Oven, Self-cleaning	14% (EF)
Gas Cooking Top	46% (EF)
Gas Oven	8.5% (EF)
Gas Oven, Self Cleaning	7.8% (EF)

Central Air Conditioners

Split System	10.3 (SEER)
Single Package	8.9 (SEER)

Furnaces

Gas, Gravity	70% (AFUE)
Gas, Forced Air	75% (AFUE)
Gas, Boilers	79% (AFUE)
Oil, Boilers	85% (AFUE)
Oil, Forced Air	82% (AFUE)
Electric	100% (AFUE)

Dishwasher

Standard	0.41 Cycles/kwhr (EF)
Compact	0.50 Cycles/kwhr (EF)

Television Sets

Color	229% (REEF)
Monochrome	176% (REEF)

Clothes Washers

Top Loading	1.01 Ft.^{3}/kwhr (EF)
Front Loading	1.25 Ft.^{3}/kwhr (EF)

Humidifiers

Central System	92% (EF)
Room	93% (EF)

Dehumidifiers

(No Classes Specified)	2.9 Pints/kwhr (EF)

Abbreviations: EF = energy factor; EER = energy efficiency ratio; AFUE = annual fuel utilization efficiency; SEER = seasonal energy efficiency ratio; REEF = receiver energy efficiency factor.

Source: U.S. Department of Energy, *Energy Efficiency Improvement Targets for Nine Types of Appliances,* F.R. 43, no. 70, Apr. 11, 1978, p. 15145.

efficiency standards and to assure the compliance of individual
units with the standards. The test procedures are flexible enough
to allow for technological variation among the different product
lines within a product type, yet are standardized enough to assure
that the different product lines will be subject to the same measurement criteria in order to provide comparable measures of energy
efficiency.

The standards are designed for achieving the maximum efficiency
improvement that is technologically feasible and economically justified. The maximum technologically feasible energy efficiency level
is, for each class of consumer products, the most energy-efficient
model that was commercially available in October 1979. The DOE's
tentative determinations of maximum technologically feasible energy
efficiency levels for each product class are listed in Table 3-16.

It should be understood that energy efficiency standards mean
performance standards, as opposed to design standards, which prescribe minimum energy efficiency levels for each unit of a covered
product. The standards do not prescribe the methods, processes, or
materials used to achieve any particular efficiency level. Energy
efficiency improvements of 20 to 30% are projected as a result of
these standards, which apply only to new products manufactured after
the effective date of the standards.

CONCLUSION

Over the years, the average number and size of energy-consuming appliances found in the home have expanded enormously. New models
which use more energy than their predecessor but which have convenience extras are being purchased. Most of the energy consumed by
appliances is utilized to heat water and to run the refrigerator and
electric range. Other appliances use very little energy, but as
the cost of energy rises, the public must be made aware of the amount
of energy required to operate these appliances.

The selection and purchase of a new appliance must be done with
care and thought, for the cost of energy, which is steadily climbing,

will take a progressively larger share of the family's budget. All aspects of energy conservation should be kept in mind when buying new equipment: Does the appliance have optional features? Are the options for convenience only, or do they also save energy? Would an additional investment in a high energy-using appliance be worth it in terms of higher energy expenses?

Newer, more efficient appliances will be available in the near future, but these appliances will not result in a reduction in energy usage unless we control the manner in which and the extent to which we use them. The opportunities available for saving energy consumed by appliances are at least as good as those for savings in heating and cooling. The key to conservation is determining what our needs are and to satisfy them at the lowest possible cost in both energy and dollars.

NOTES AND REFERENCES

1. John V. Fechter, *Human Factors in Appliance Energy Consumption,* National Bureau of Standards, May 1977, pp. 2-7.

2. Robert A. Hoskins and Eric Hirst, *Energy and Cost Analysis of Residential Water Heaters,* ORNL/CON-10, Oak Ridge National Laboratory, Oak Ridge, Tenn., June 1977.

3. National Bureau of Standards, *Appliance Energy Efficiency Improvement Target for Clothes Washers,* U.S. Department of Energy, 1977, pp. 16-17.

4. Energy Research and Development Administration, *Insulation Refit Kit for Domestic Water Heaters,* U.S. Department of Energy, Mar. 23, 1977, p. 108.

5. Arthur D. Little, Inc., *Study of Energy Saving Options for Refrigerators and Water Heaters,* vol. I, Refrigerators, Federal Energy Administration, May 1977.

6. Dennis O'Neal, Janet Carney, and Eric Hirst, *Regional Analysis of Residential Water Heating Options: Energy Use and Economics,* ORNL/CON-31, Oak Ridge National Laboratory, Oak Ridge, Tenn., October 1978.

7. Science Applications, Inc., *Appliance Efficiency Program Final Report,* no. SAI 76-551-LJ, LaJolla, Calif., 1979, pp. 3-36.

8. U.S. Department of Energy, *Appliance Energy Efficiency Improvement Target for Clothes Dryers,* 1977, pp. 21-23.

9. W. David Lee, W. Thompson, X. Lawrence, and Robert P. Wilson, *Design, Development, and Demonstration of a Promising Integrated Appliance*, Arthur D. Little, Inc., Cambridge, Mass., September 1977.

4

Solar Heating/Cooling/Hot Water

Melvin H. Chiogioji
U.S. Department of Energy
Washington, D.C.

Most of the energy available on earth is created by the sun: all
organic fuels, including fossil fuels, originate from plant life
which need sunlight in order to survive, and winds result from the
uneven heating of the earth's surface and atmosphere by the sun.

The energy from the sun is transmitted by electromagnetic radiation. Forty-six percent of the radiation that reaches the earth is
visible light. Invisible, shorter wavelength radiation, called
ultraviolet radiation, comprises 4.6%. Most of this short wavelength
radiation is absorbed by the earth's atmosphere; very little actually
strikes the earth's surface. The greatest portion (49%) of the
radiation that reaches the earth is of longer wavelength than that
of visible light and is called infrared radiation, or heat energy.
The remainder of the radiation from the sun consists of ionizing
ultraviolet and x-ray radiation.

Although a constant amount of solar energy strikes the outer
atmosphere (429 Btu ft^2/hr), the amount of energy available on the
earth's surface is a function of many variables. Precipitation,
clouds, fog, smog, airborne dust, incidence angle of the sun, and
the length of the path that the sun's rays must travel through the

earth's atmosphere all determine the quantity and quality of sunlight a particular area receives. Theoretically, if the solar
energy of the outer atmosphere could be collected 24 hr/day for 30
days, it would amount to approximately 1000 kWh/month/m^2 (3.4 million Btu/month/ft^2). Just 2 m^2 (21.4 ft^2) of this energy would be
needed to provide all the energy that a typical residence requires
for heating, cooling, and hot water (1, p. 53).

Since the United States is in the midst of a serious ongoing
energy crisis, new ways to meet our growing energy demands must be
found. One possibility that has lately been receiving a great deal
of attention is that of capturing and using the sun's radiant energy.
Although there are several ways in which the sun's rays can be har-
nessed, this chapter primarily discusses the direct transfer of
solar energy into heat, particularly for space and hot water heating
and for space cooling. We seek to provide building and home owners
with a basic understanding of solar heating, cooling, and hot water
systems--what they are, how they perform, the energy savings possible, and the cost factors involved--with data reflecting the performances of several existing systems.

PASSIVE SOLAR DESIGN

Because of its large heat storage capacity, the earth takes a long
time to cool off after the sun goes down and a long time to warm up
after the sun rises. This accounts for afternoon temperatures being
higher than morning temperatures, in spite of the fact that equivalent amounts of solar radiation are available at both times. The
same principle accounts for the time lag between the earth's and
sun's seasons. While midsummer for the sun is around June 21 in the
northern hemisphere, the warmest weather in the same hemisphere
usually occurs in July and August.

Building design should be based on similar principles. A building should not have wide temperature variations from one hour to the
next or from cold nights to warm days. It should not even be affected by the wider temperature extremes of summer and winter. The

adobe home of the Pueblo Indians in the southwestern United States
is an architectural example which fulfills these criteria. During
the day the thick walls of clay store the heat of the sun, preventing it from reaching the interior of the home. At night when the
temperature falls, the stored heat warms the interior. Conversely
the cool air stored in the walls during the night keeps the home
cool throughout the day.

In addition to eliminating the effects of daily temperature
variations, energy-conscious design should enable a building to
modulate itself so that the effects of seasonal temperature extremes
are minimized. It should be possible to open up the building on
sunny winter days so that the sun can shine in, and then to "button"
the building up tightly at night to keep that heat from escaping.
Conversely, during the summer it should be possible to close up the
building against the heat of the day and open it up again at night
to receive the cooler air.

The most effective way to use the sun for heating then is to
design and use a building in such a way as to avoid a reliance on
mechanical systems. A building must satisfy three basic requirements in order to achieve this (2).

1. *The building must be a solar collector.* It must let the sun in
   when heat is needed, and it must keep it out when a cooler temperature is desired. This is done primarily by designing the
   building so that the sun can penetrate through the walls and
   windows during the winter and, by using shading devices such as
   trees, awnings, and venetian blinds, keeping the sunshine out
   during the summer.

2. *The building must be a solar storehouse.* It must be able to
   store the heat for cool (and cold) times when the sun is not
   shining, and to store the cool for warm (and hot) periods when
   the sun is shining. Buildings which are built of heavy materials,
   such as stone and concrete, do this most effectively.

3. *The building must be a good heat trap.* It must be able to make
   good use of the heat (or cool) and to let it escape only very

slowly. This can be done by using insulation, reducing air infiltration, and installing storm windows.

The concept of using sunlight and winds to help heat and cool
houses is rapidly being "rediscovered" in the wake of today's new
consciousness about the use of energy. Features that were standard
construction practices before the introduction of central heating
and cooling systems are making a comeback. Many of the houses being
built today use natural phenomena to minimize their use of conventional fuels for heating and cooling. Overhangs are used to admit
sunlight in the winter and provide shade in the summer. Deciduous
trees ensure summer shade and winter sunshine, and windbreaks tem-
per the effects of winter winds.

Direct-Gain Systems

The simplest passive solar design incorporates the addition of larger south-facing windows to a house. But covering the south wall
with too many windows could provide the typical frame house with too
much heat, even in cold weather, because of the house's limited
capacity to absorb and store heat. Adding more windows necessitates
the addition of massive features to the construction of the house
so that more heat can be stored, but with only small increases in
the interior temperature.

Such direct-gain systems are the ultimate in simplicity, as
they use no fans or dampers and the heat "stays" when the electricity goes off. However, the large windows can produce uncomfortable
glare, and if the floor is to be effectively used to store heat, it
must not be carpeted. If the house does not incorporate massive
components very effectively, the resulting temperature swings can
be uncomfortably large--as much as 20 to 25°F in a day.

Indirect-Gain Passive Buildings

The problems with glare and the wide temperature fluctuations experienced in many direct-gain buildings have led to a variety of
simple approaches in which the sun does not directly heat the

living space. Several of these approaches are discussed in this
section.

Among the most widely publicized solar houses are those being
built in France by Felix Trombe and Jacques Michel. The Trombe design is characterized by a heavy, monolithic wall which faces the
south. It is single- or double-glazed, with an airspace about 4 in.
wide between the sun-facing surface of the concrete and the inner
glass. Small openings at the bottom of the Trombe wall admit air
from near the floor of the residence into the airspace in the wall.
When the sun shines on the concrete wall, the air is warmed; it
rises and reenters the house through openings near the ceiling. As
the exterior surface of the concrete becomes very hot during the
daylight hours, heat begins to flow inward, eventually warming the
entire mass of concrete. This heat, stored in the Trombe wall,
radiates into the living space as heat is required.

A house erected by Steve Baer in 1974 near Albuquerque, New
Mexico uses 55-gal drums filled with water and laid on their sides
to absorb incoming solar radiation, store it, and transfer it as
needed into the residence (3, pp. 166-168). The south-facing wall
adjacent to the outer ends of the water drums is glazed with a
single $\frac{1}{8}$-in. glass. An insulating and reflective panel is mounted
on the outside of the glass, so that it can be lowered during the
daylight hours to reflect additional solar radiation onto the drum
collectors, and raised again at night to minimize loss of heat outward. This system has operated successfully, with a minimum of
maintenance.

A variation of the mass wall approach is the "greenhouse" or
"attached sunspace" design. Sunlight provides all of the heat for
the sunspace, which is a large, live-in collector, and part of the
heat for the rest of the house. The storage wall is placed between
the sunspace and the living quarters. The storage wall mass evens
out the temperature variations for the living quarters, while the
temperature in the sunspace undergoes wide swings.

Hybrid Passive Systems

The definition of passive systems is the object of some controversy.
The term "passive" implies an absense of machinery and controls, but
some houses incorporate a storage bed to improve the system's performance and fans and controls to circulate heat to and from the
storage bed. For example, the Balcomb house provides 95% of its
heating needs by using a greenhouse in combination with a rock storage bed. The wall between the greenhouse and the living quarters
is a heavy adobe wall which stores heat during the day and releases
it to the living space as it is needed. Whenever the temperature
in the greenhouse rises above a certain point, blowers force air
through the rock storage bed to store the heat that is used to warm
the house after the adobe has exhausted its heat supply. This house
is near Santa Fe, New Mexico, where subzero winter temperatures occur, yet it requires less than 1000 kWh of electricity for supplementary heating (the equivalent of 30 to 40 gal of heating oil during a 6400 degree-day winter) (4).

Cost of Heat From Passive Solar Heating Systems

A variety of passive heating systems are capable of providing substantial heat to buildings. These systems are simple and can be
built from ordinary construction materials. As with other new systems, the installation costs and the values of the energy savings
will be major factors in general consumer acceptance.

The cost of heat from passive solar heating systems depends on
the system itself and the costing methodology used. The most comprehensive survey now available of the cost and performance of passive solar heating installations is the one performed by Deborah
Buchanon of the Solar Energy Research Institute (5). It presents
information on installation costs and amounts of useful heat delivered based on a study of 32 actual systems and 19 simulated installations. The heat delivered by 20 of the systems was determined
by monitoring the building performance, and the heat delivered to
the other buildings was determined through engineering estimates.

Table 4-1 presents the cost of delivered heat for the systems surveyed. The cost of delivered heat was calculated by using an annual capital charge rate of 0.094 and an annual maintenance estimate of 0.005 of the initial system cost, based on the average maintenance cost in the survey. It is assumed that the systems have no intrinsic value other than as heating systems.

Summary

Passive solar buildings, which can obtain 20 to 95% of their required heat from the sun without the help of active solar collectors, have been built and are operating in many parts of the country. While data are sketchy, it appears that such housing is competitive with electric-resistance heating in terms of cost, and is even competitive with oil and gas in some parts of the country. In addition to supplying heat, many passive solar homes provide additional space, good natural light, and a pleasing view; these characteristics may be as important to homeowners as the savings in fuel costs. An additional bonus is that the systems are simple constructed, with few moving parts, so that little maintenance is required.

Table 4-1. Cost of Heat from Passive Solar Heating Installations

Installation Type	Cost of Heat ($/Million Btu)
Direct Gain:	
Monitored..........................	3.00-13.80
Unmonitored	3.30-12.60
Thermal Storage Wall:	
Monitored..........................	3.60-24.20
Unmonitored	8.80-15.40
Thermal Storage Roof:	
Monitored..........................	8.20
Greenhouse/Attached Sunspace:	
Unmonitored	1.20-11.90
Hybrid:	
Monitored..........................	2.60-16.60
Unmonitored	11.90

Source: Based on Ref. 5.

ACTIVE SOLAR SYSTEMS

The alternative to a passive solar system is an active system, which
employs mechanical means to move the sun's energy. Either liquid
or air is used to collect and transport the sun's energy, and pumps
and/or fans are used to move the liquid and/or air within the system. Active solar energy systems can be used to heat water for domestic use, industrial processes, and swimming pools. They can also
provide energy for heating buildings and to power equipment for
cooling.

Figure 4-1 is a schematic illustration of an active solar heating and cooling system, representative of those available today.
Active systems currently consist of the following five units essential to the collection, control, and distribution of solar heat:

1. *The solar collector,* which intercepts solar radiation and con-
   verts it to heat for transfer to a thermal storage unit or to
   the heating load.

2. *The thermal storage unit,* which allows heat to be stored for
   later use at night or on cloudy days if more heat is collected
   than is required for space heating or domestic hot water heating. This thermal storage material can be either liquid, rock,
   or phase-change material. Storage units generally operate independently of any cooling or heating requirement, since they
   collect and store solar energy whenever there is sufficient
   solar radiation.

3. *An auxiliary heat source,* which is required as a backup to the
   heating/cooling system when the solar system is unable to meet
   heating/cooling demands. Although the solar collector can be
   sized large enough to provide the full heating load throughout
   the year, doing so is not as economical as using an auxiliary-
   assisted system. A solar system which is sized to handle 50 to
   80% of the average load may be able to handle the entire load,
   but on cloudy days the conventional heating system may have to
   handle the entire load.

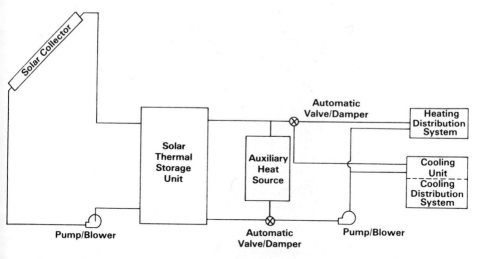

Figure 4-1. Schematic of a solar heating and cooling system.

4. *A heat distribution system.* Heat distribution to the building can be accomplished in a number of ways. In an air system, the solar-heated air can be taken directly from either the solar collector or the thermal storage unit and delivered throughout the building by utilizing a blower and duct distribution. In a liquid system, a liquid-to-air heat exchanger can be used to provide heat for a central air distribution system, or the liquid can be piped directly to the heated space, where separate fan-coil units can be used to heat the building.

5. *A cooling unit.* A variety of devices can be used for cooling, among them, absorption cooling units (both lithium-bromide and ammonia-water systems) and Rankine cycle vapor-compression units. However, only the lithium-bromide absorption unit is commercially available, and it has been used only in experimental installations. A heat pump might be utilized as a conventional cooling unit (powered by electricity) and used as the auxiliary for the solar space heating system.

Figure 4-2 is a schematic drawing of a domestic hot water system (DHW) which utilizes solar heat. Solar energy from a collector

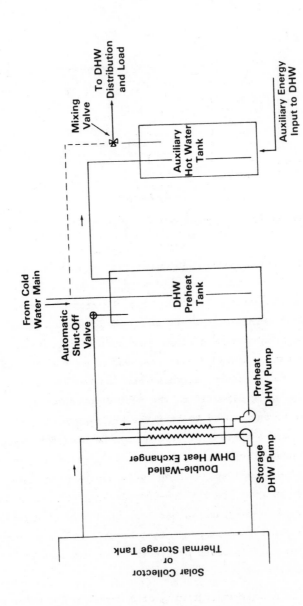

Figure 4-2. Schematic of a domestic hot water system.

or a thermal storage unit is used to preheat the water from the cold water main. This preheated water replaces the hot water taken out of the auxiliary hot water tank as hot water is used in the building. Conventional fuels, such as gas or electricity, are used to boost the temperature of the preheated water to the desired temperature (e.g., 140°F), and/or to keep the water remaining in the auxiliary tank at the desired temperature.

Although both air and liquid can be used to collect and transport the sun's energy, there are significant differences in the respective systems. Air systems are more effective in heating when the warm air from the collectors can be blown directly into the space to be heated. Air systems can also be used for preheating water but have a limited capacity for cooling. Liquid systems are widely used for space heating and for the heating of water. Moreover, they can provide energy at high enough temperatures to power chillers for space cooling.

Air systems are simpler and cost less, but operate at lower efficiency and take up more space than do the more complicated liquid systems. Dust and mildew accumulations in the storage bins of air systems can create long-term operational problems. Liquid systems, on the other hand, have problems with clogging, corrosion, and freezing. Leaks in the ductwork of air systems reduce operational efficiency but are not consequential as are leaks in liquid systems, which can cause system breakdowns. Furthermore, leaks in liquid systems used for heating potable water can result in pollution of the water.

SOLAR COLLECTORS

There are several types of solar collectors (1, pp. 120-125). These are devices that convert incident solar radiation to useful energy, usually in the form of heated air or heated liquid (see Table 4-2). Nonfocusing collectors absorb the radiation that falls directly on a flat absorber plate, but the temperatures that these collectors can provide are limited. Other collectors, although designated flat-

Table 4-2. Types of Collectors

Collector Type and Nomenclature		Medium	Temperature Range
● Nonfocusing	Flat Plate	Liquid or Air	70°F to 180°F
● Partial Focusing	Flat Plate, Evacuated Tube	Liquid	180°F to 250°F
(Note: All of Above may be Augmented by Reflectors)			
● Full Focusing Tracking Concentrators			
Linear Focus	Tracking Parabolic Reflector with Fixed Absorber	Liquid	
	Tracking Absorber with Fixed Parabolic Reflector	Liquid	180°F to 300°F
	Fresnel Lens Tracking Assembly	Liquid	
Point Focus	Tracking Paraboloid Concentrator	Liquid	Above 300°F

Source: Ref. 1, p. 120.

plate collectors, employ curved internal reflectors which partially focus the sun's energy on absorber tubes. These partial-focusing collectors can attain higher temperatures than can nonfocusing collectors. External reflectors are often used as adjuncts to both the nonfocusing and partial-focusing collectors, in order to increase the amount of radiation falling on their surfaces.

When higher temperatures are desired, full-focusing tracking concentrators which follow the sun are used. Some track on a single axis, some on two axes. Some focus linearly on absorber tubes, while others focus on a small point, which can provide very high temperatures. The most common linear-focusing collector is a tracking parabolic reflector with a fixed absorber tube mounted above it. The absorber tube remains stationary as the reflector tracks the sun on one axis. A second type of linear-focusing collector is one that employs a fixed parabolic reflector in conjunction with an overhead absorber tube which moves (on one axis) so as to stay within

the focal plane of the reflector. Still another linear-focusing
collector uses a single axis tracking lens/absorber assembly. A
tracking paraboloid concentrator, which normally tracks on two axes,
is used for point focus.

Liquids may be used as the collection and transfer medium in
all of these collectors. Air is normally used only in the nonfocusing flat-plate collectors. A typical air collector panel consists of a frame containing transparent glazing to admit radiant
energy from the sun, a dead airspace to provide insulation, an absorber plate to collect and transmit the radiant energy, an open
space through which air is circulated to remove the collected energy, and back and side insulation to minimize thermal losses (see
Fig. 4-3).

The absorber plate, which collects the solar energy, is normally painted black. It may have a selective coating to maximize
the amount of heat absorbed and minimize that which is reradiated.

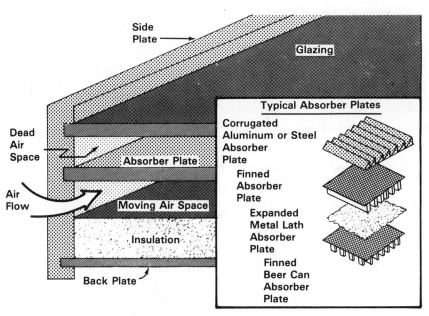

Figure 4-3. Air collector panel.

Typical absorber plates also have baffles that extend into the moving air cavity to create turbulent flow for increasing heat transfer.
The air collector panels also contain connections for assembly into
collector arrays. The sizes of the panels vary among manufacturers
but are normally of such dimension and weight that they can be handled by two men.

A typical flat-plate liquid collector consists of one or more
transparent glazings to admit the sun's energy (see Fig. 4-4).
These glazings are separated by dead airspace. The radiation which
passes through the glazing falls on and heats an absorber plate that
is painted black or with a selective coating to increase the amount
of energy it collects and retains. The absorber plate is usually
made from copper or aluminum; the conduits within it, through which
liquid is circulated to collect and transport the sun's energy, are
made up of attached tubes or are formed by means of corrugations,
dimples, or a quilted pattern in two adjoining plates. The collector panel is insulated at the back and on the edges to minimize heat
losses.

A trickle-down collector is the simplest of the flat-plate collectors (Fig. 4-5). It consists of an upper manifold which has
spaced openings that allow water to trickle into the flutes of a
corrugated metal absorber plate. The water is heated as it flows
over the hot metal and is caught in a gutter at the bottom, to be
returned to storage. Such a collector may or may not have glazing.
Backside insulation minimizes heat losses. This type of collector
lends itself easily to on-site construction.

An evacuated tube collector panel consists of several evacuated
tubes, mounted in parallel and connected with internal manifolds in
a flat panel (Fig. 4-6). Integral curved reflectors mounted beneath the tubes focus the sun's energy toward the centers of the
tubes. Liquid enters through a feeder tube, travels to the far end
of the tube assembly, reverses, and returns through a larger concentric tube. The large concentric tube has a selective coating on the
outside to maximize the energy it receives and minimize the energy

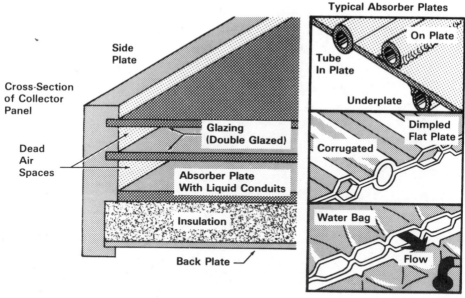

Figure 4-4. Flat-plate liquid collector panel.

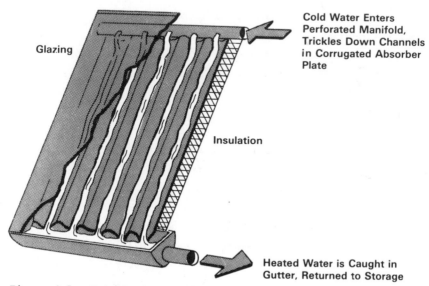

Figure 4-5. Trickle-down collector.

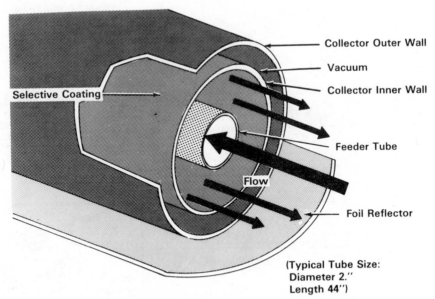

Figure 4-6. Evacuated tube collector (several tubes mounted in
parallel to form collector panel).

it rejects or reradiates. This tube is separated from the larger
outer transparent glass tube, by means of a hard vacuum which provides the maximum amount of insulation.

Many factors affect collector performance. For example, in
the northern hemisphere, collectors should face a southerly direction, preferably due south. Topography, vegetation, and adjacent
structures may limit the amount of energy collected, because of
shading. Wind will increase the convection losses from the collector surface, and condensed moisture affects the transparency of the
collector glazing. Moisture also affects the ratio of diffuse to
direct radiation. Finally, the amount of available energy is dependent on the climate of the region.

Figure 4-7 illustrates the effects of the sun angle and of the
collector tilt angle on the amount of energy available to the collector at 40° north latitude. As would be expected, horizontal collectors are most effective in the summer months, when the angle of

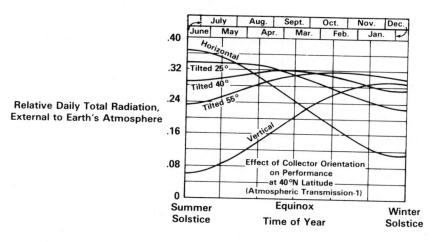

Figure 4-7. Yearly performance based on collector orientation. (From Ref. 1, p. 60.)

inclination of the sun is the greatest; however, horizontal collectors are not very effective in the cooler months. Vertical collectors, conversely, are most effective in the winter and least effective in the summer. Tilt angles between these extremes (the tilt angles shown are measured from the horizontal) offer better year-round performance. For this latitude, an optimal tilt angle for the collectors, if they are to be used year round, is approximately 40°. If the collectors are only to be used for heating, a 55° tilt angle appears more desirable. It should be noted that at this latitude, a vertical collector has good performance during the winter, which makes attractive the use of conventional windows for gathering the sun's energy in passive heating systems.

THERMAL STORAGE

Because of variations in solar radiation and atmospheric temperatures and because of the resulting noncorrespondence between available solar heating and heating load demand, some form of energy storage is required. This energy storage requirement can be met most economically by some form of thermal storage, that is, storage

of heat (or cool) for heating and cooling systems. The types of
thermal storage units which might be utilized are quite varied, but
common storage mediums for solar systems are water, rock aggregate,
and materials which undergo some sort of phase change. Water is
used for storage in systems that employ liquid for the collector
medium, rock aggregate is normally used in systems that use air, and
phase-change materials are used with both liquid and air systems.

In most liquid systems, water that is used as the storage medium is contained in insulated tanks made of steel, concrete, or
fiberglass. The storage water may be the same water that is circulated through the collectors or may be separated from the collector
fluid by a heat exchanger. The water from the storage tank is normally circulated through external heat exchangers to transfer the
stored heat wherever it is needed. In some instances, heat exchangers are immersed into the storage tank to absorb heat for the load.

Rock aggregate of pebble beds, which are used for heat storage
in solar systems that use air as the collector medium, are usually
located in a large, externally insulated concrete bin. Hot air from
the collectors enters an open space at the top of the bin and is
warmed as it is drawn through the aggregate. The air then passes
through an open space at the bottom of the bin and back through the
collectors. In the heating-from-storage mode, hot air is withdrawn
from the top of the bin and blown into the bottom and circulates
back through the aggregate where it is warmed for a subsequent cycle.

Most phase-change materials used for thermal storage are in the
solid state, but as energy is stored in the material, it liquifies
[phase-change materials normally have a melting point (freezing
point) of 80^{0}F or better].

Glauber salt, which melts at approximately 100^{0}F and absorbs
104 Btu/lb as it melts, is one of the more common phase-change materials used for storage. However, because of the low conductivity
of phase-change materials in a solid state, it is difficult for heat
to be absorbed and distributed at a fast enough rate to furnish the
amount of heat required. Consequently, phase-change storage systems

must have many separate compartments, making the ratio of surface
area to size of any frozen piece relatively large. Phase-change
storage systems resemble ice trays stacked in a refrigerator, with
spaces between them for the air of liquid transfer medium to circulate.

In the collector-to-storage mode, hot air is circulated around
these trays or cylinders and returned to the collectors. As heat is
absorbed from this circulating air, the salts melt. This change in
state, which takes place at a constant temperature of 80°F or above,
absorbs considerable energy in the process. In the heating-from-
storage mode, air circulates over the cylinders or trays to absorb
the stored heat. As heat is withdrawn from the salts, they return
to the solid state.

Figure 4-8 shows the relative efficiencies of the three types
of storage systems. As is shown, rock requires the greatest amount
of volume to store a given amount of energy because of its relatively

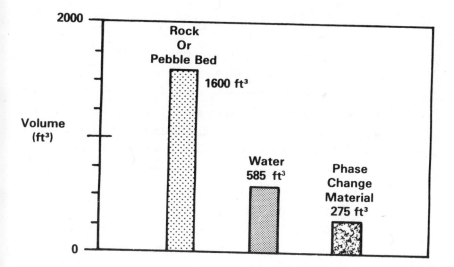

Figure 4-8. Storage subsystems. Volume of storage required for 5
million Btu. Note: Storage required to maintain 70°F inside temperature for 3 days for typical residence with outside temperature
of 20°F (storage temperature = 200°F). (From Ref. 1, p. 145.)

low thermal capacity (specific heat) and because of the spaces that
are required for the air to circulate between the rocks. Technical
difficulties and economic disadvantages are the reason phase-change
storage materials are not yet ready for practical use in solar heating and cooling systems. Thus, most systems concentrate on hot
water and pebble-bed thermal storage units.

SOLAR SPACE HEATING SYSTEMS

Collecting, storing, and using solar energy for space heating requires the control of the air or liquid flow, or both. Three modes
of operation are possible:

1. Heating the building directly from the collectors
2. Heating the storage unit from the collectors
3. Heating the building from storage and/or auxiliary heating units

Air Systems

Figure 4-9 shows an arrangement for heating a building directly from
the collectors of an air system. This mode would be used whenever
the sensor in the discharge air stream of the solar collector indicates that the collector temperature exceeds a predetermined point,
and the room thermostat still demands heat.

The duct arrangement shown depicts a fairly typical two-blower
system, in which dampers J-1 and J-2 are positioned to isolate the
storage circuit, damper J-3 is left opened, and the economizer is
closed. The airflow path follows the shaded area, provided collector flow rate and auxiliary heating airflow requirements are the
same. Whenever there are differences between collector flow rates
and airflow requirements, the bypass damper J-4 is opened to allow
some portion of the airflow to circulate through the bypass duct.
Both furnaces fan and collector fan must operate in this mode.

When the collector air temperature is below a predetermined
set point and the storage temperature is above a specific point
(e.g., 90°F or above), dampers J-1 and J-2 must be repositioned to

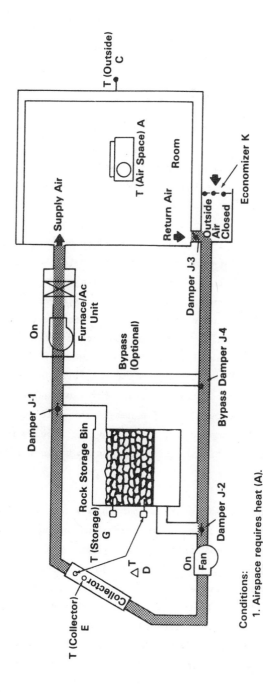

Figure 4-9. Heating directly from collectors of an all-air system. (From Ref. 6, p. 7-1.)

direct the air through the pebble-bed storage unit. At this time,
the collector fan would shut off (see Fig. 4-10). In most instances,
when the temperature in the pebble-bed storage drops and there is
insufficient heat to satisfy space heating needs, the drop in space
temperature will activate the auxiliary heater. This unit may be a
gas, oil, or electric furnace, or perhaps a heat pump.

When there is no demand for heating but there is heat available
at the collector, the system switches to the heat storage mode, as
shown in Fig. 4-11. In this mode, dampers J-1 and J-2 are positioned
to route the discharge air from the collector through the pebble-bed
storage. The collector fan turns on, and whether or not the auxiliary furnace fan turns on is dependent on whether continuous air circulation or cycle fan operation is preferred. Note that the flow
path for storage heat is the reverse of the path used to remove heat
in the heating-from-storage operating mode. This reverse flow-path
direction is essential in order to take full advantage of the temperature stratification which occurs in the rock storage and to
utilize the hottest air when heating from storage.

In a typical operation, the collector fan turns on whenever
the temperature difference between the air entering and leaving the
collector exceeds 20°, regardless of a demand for heat from the room
thermostat. Dampers automatically adjust for the storage mode, and
heat storage continues until the room thermostat calls for heat.

Liquid Systems

Liquid solar heating systems operate in much the same way as air
systems (6, pp. 7-6 to 7-8), the only difference being the addition
of a liquid-towater heat exchanger in the system.

Figure 4-12 shows space heating as it is provided from the collectors when the room thermostat has indicated a need for heat and
the temperature of the collector liquid is above some predetermined
point (e.g., 90 F or higher). The collector pump and storage pump
are turned on, and the liquid/waterflow follows the path through the
duct coil. The furnace or air handler blower is also turned on to
circulate room air over the hot duct coil.

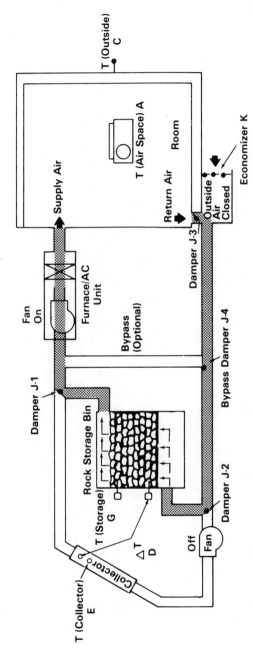

Figure 4-10. Space heating from storage. (From Ref. 6, p. 7-2.)

Conditions:

1. Airspace requires heat (A).
2. Collector temperature (E) is below control point.
3. Storage temperature (G) is above control point.

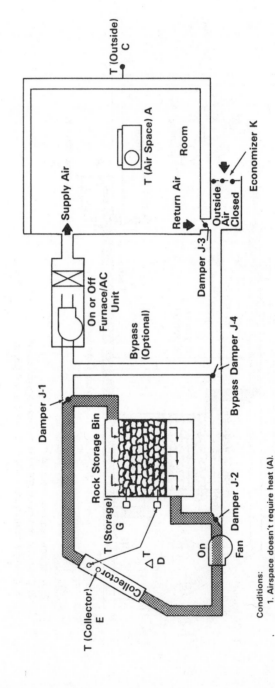

Figure 4-11. Storing heat. (From Ref. 6, p. 7-3.)

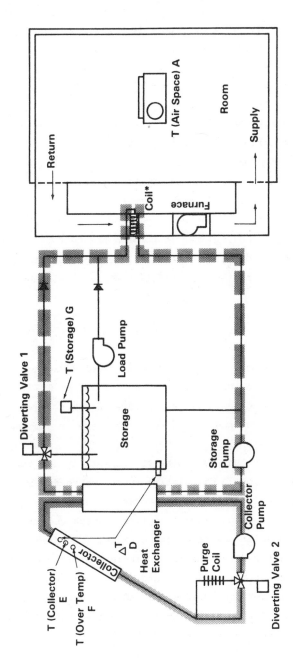

Figure 4-12. Heating from the collector. (From Ref. 6, p. 7-7.)

*Note: If heat pump is used, solar coil is placed downstream from heat pump coil.

Conditions:
1. Airspace requires heat (A).
2. Collector temperature (E) exceeds control point.

Figure 4-13 shows a situation in which the room still requires heating but the collector liquid temperature is below the control point, while the storage temperature exceeds the predetermined temperature, again, perhaps $90^{\circ}F$. Both storage and collector pumps are stopped, and the load pump is started in order to take heat from storage and circulate it through the duct coil.

Figure 4-14 illustrates the heat-storing mode. When space heating requirements are satisfied, and the temperature differential between the collector and storage is high enough (e.g., $20^{\circ}F$), excess heat can be stored in the storage unit. Diverting valve 1 is activated to divert the discharge from the heat exchanger into the top of the storage tank. The storage pump removes cooler water at the bottom of the tank and recirculates it through the heat exchanger so that the temperature of the water in the storage unit is raised. Thus, this heated water can be used to fulfill heating requirements at a later time.

SOLAR WATER HEATING

Solar water heating is the most widely used form of solar energy in the residential sector in the United States. By the end of 1977, an estimated 63,000 homes had installed solar water heaters (new and retrofit), while only 3000 had installed solar space or combined solar space and water heaters (7).

The solar water heaters that have been experimentally and commercially used are of two types: circulating types, which supply solar heat to a fluid circulating through a collector and store hot water in a separate tank; and noncirculating types, which use water tanks that serve as both solar collector and storage. The circulating water heaters are either direct heating, single-fluid types, in which the water is heated in the collector and storage or by pumped circulation between collector and storage; or indirect heating, dual-fluid types, in which a nonfreezing medium is circulated through the collector for subsequent heat exchange with water.

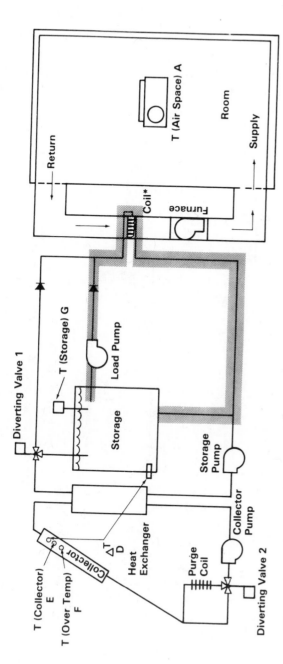

*Note: If heat pump is used, solar coil is placed downstream from heat pump coil.

Conditions:
1. Airspace requires heat (A).
2. Collector temperature (E) is below control point.
3. Storage temperature (G) exceeds control point.

Figure 4-13. Heating from storage. (From Ref. 6, p. 7-8.)

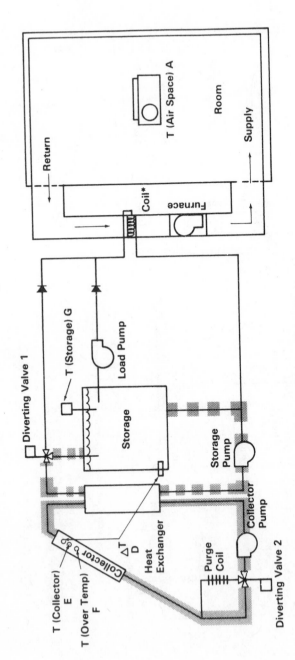

Figure 4-14. Heat storage mode.   (From Ref. 6, p. 7-8.)

Direct Heating, Thermosiphon Circulating Type

The most common type of solar water heater, used almost exclusively in nonfreezing climates, is shown in Fig. 4-15. The thermosiphon system, which is the simplest solar water heating system, combines a flat-plate collector with a storage tank mounted above the collector. The tank is placed high enough for cold water in the downcomer tube to displace, by convection, the hot water in the collector; this causes a slow circulation of water. The collector is usually single-glazed and may vary in size from about 30 to 80 ft^2. The insulated storage tank is commonly in the range of 40- to 80-gal capacity, which will generally meet the hot water requirements of a family of four persons living in a sunny climate.

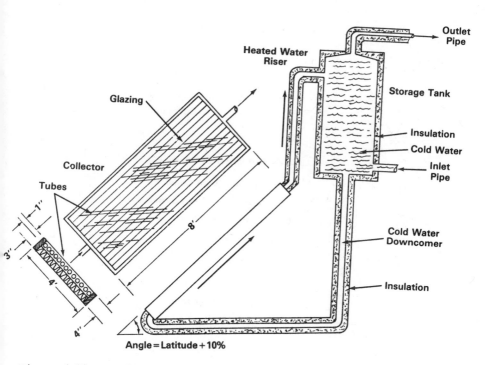

Figure 4-15. Typical thermosiphon solar water collector.

Direct Heating, Pump Circulation Type

If placement of the storage tank above the collector is inconvenient
or impossible, the tank may be located between the collector, with
a small pump used for circulating water between the collector and
the storage tank. A schematic arrangement is shown in Fig. 4-16.
This arrangement is usually more practical in the United States than
the thermosiphon type, because the collector is usually located on
the roof of a building, with a storage tank in the basement. On
sunny days, a temperature sensor actuates a small pump which circulates water through the collector-storage loop for heat storage.

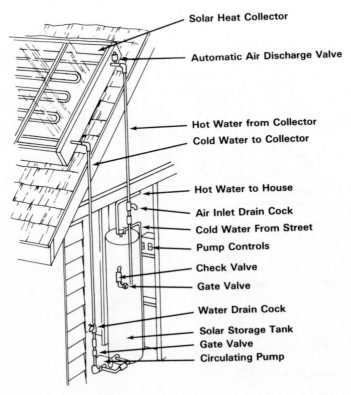

Solar Heat Collector

Automatic Air Discharge Valve

Hot Water from Collector
Cold Water to Collector

Hot Water to House

Air Inlet Drain Cock
Cold Water From Street
Pump Controls

Check Valve

Gate Valve

Water Drain Cock

Solar Storage Tank
Gate Valve
Circulating Pump

Figure 4-16. *Roof-mounted pumped solar water heating system.*

Direct Heating, Pump Circulation, Drainable Type

If the solar water heater is used in a cold climate, it may be protected from freeze damage by draining the collector when subfreezing temperatures occur. Drainage of the collector can be accomplished by automatic valves which route water outflow to a drain (sewer) and allow the inflow of air to the collector.

An alternative method for draining the collector is based on the use of a nonpressurized collector and storage assembly. In this system, a float valve in the storage tank control controls the admission of cold water to the tank, and a pump in the hot water distribution system furnishes the necessary water pressure. The solar collector will drain into the storage tank whenever the pump is not operating, and air enters the collector through a vent to prevent freezing.

Circulating Type, Indirect Heating

The drainage requirement during freezing weather can be eliminated by using both a nonfreezing heat transfer medium in the solar collector and a heat exchanger (inside the building) for transferring heat from the solar heat-collecting medium to the hot water tank. Figure 4-17 illustrates a method for solar water heating which uses a liquid heat transfer medium in the solar collector. The most commonly used liquid is a solution of ethylene glycol (which is automobile radiator antifreeze) in water. A pump circulates this unpressurized solution, as is done in the direct water heating system, and delivers the liquid to and through a liquid-to-liquid heat exchanger. Simultaneously, another pump circulates domestic water from the storage tank through the exchanger, back to storage. The control system is essentially the same as that in the design which utilizes water in the collector.

Water Heater Sizing

Seasonal variations can result in large differences in the availability of solar energy for a hot water system. In the winter, for example, an average recovery of 40% of an available 1200 Btu of solar

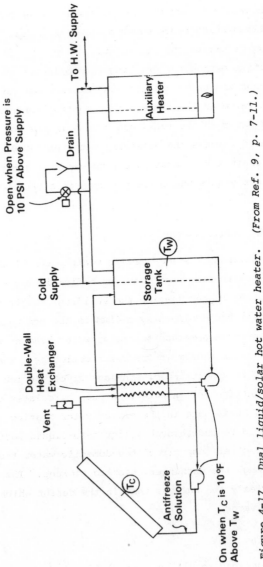

Figure 4-17. Dual liquid/solar hot water heater. (From Ref. 9, p. 7-11.)

energy per square foot would require approximately 100 ft^2 of collector for a 50,000-Btu daily heating requirement (8). Such a design could provide essentially all of the hot water needs of an average winter day but would fall short on days of less than average sunshine. By contrast, a 50% recovery of an average summer solar supply of 2000 Btu/ft^2 would require only 50 ft^2 of collector to satisfy the same hot water requirements.

A 50-ft^2 collector could supply the major part of summer hot water requirements but would probably supply less than half of winter needs. If a 100-ft^2 collector were employed so that winter needs could be more nearly met, the ysystem would be oversized for summer operation, and the excess solar heat would be wasted. The more important disadvantage of the oversized collector (for summer operation) is the economic penalty associated with investing in a collector that is not fully utilized. Although the cost of the 100-ft^2 collector would be approximately double that of the 50-ft^2 collector, its annual useful heat delivery would be considerably less than double. It would, of course, deliver about twice as much heat in the winter deason, when nearly all of it could be used, but in the other seasons, particularly in summer, heat overflow would occur. The net effect of these factors is a lower economic return, per unit of investment, by the larger system.

Thus, practical design of solar water heaters should be based on desired hot water output in the sunniest months, rather than at some other time of year. If based on average daily radiation in the sunniest months, the unit will be slightly undersized on partly cloudy days during the season, making necessary some form of auxiliary heat.

SOLAR COOLING SYSTEMS

There are three active solar cooling systems: evaporative coolers, absorption chillers, and solar/Rankine power chillers.

Evaporative coolers, which have found widespread use in areas of low humidity, are the oldest known solar cooling systems. The

performance of these systems is dependett on relative humidity: The
lower the humidity, the more effective the system. In areas where
relative humidity is variable, the performance of such units is
sporadic.

The recent emphasis on the use of solar energy for space heating and cooling has resulted in absorption chillers becoming the
primary equipment for space cooling. An absorption chiller uses
heat as the power source for a thermal cycle which removes heat at
the evaporator to provide cooling. Figure 4-18 shows a schematic
drawing of an absorption air conditioner. A solution of lithium
bromide and water is pumped to a generator, to which heat is provided by the solar energy system. The heat releases water vapor
from the lithium bromide solution, and this water vapor passes to a
condenser, where it is cooled and condensed by water from a cooling
tower. The water from condensation then moves through an expansion
valve into an evaporation chamber, which operates at a lower pressure. As the water evaporates, its phase change absorbs heat and
cools the fluid circulating in a heat exchanger adjacent to the
evaporator. The chilled fluid in the heat exchanger is then piped
either to a cold storage tank or to heat exchangers for direct space
cooling. The water vapor from the evaporator is piped through and
absorbed by a highly concentrated lithium bromide solution. The
resulting solution is then returned to the generator for another
cycle.

The absorption shillder cycle requires very little power to
function internally, but large amount of parasitic energy are required to pump the hot water to the generator, to pump the cooling
water from the cooling tower to the condenser, and to pump the water
to be chilled to the evaporator. Optimally, this cycle will provide
650 Btu of cooling for each 1000 Btu of solar energy provided at the
generator (1, pp. 7-11 to 7-14).

The sun's energy can also be used to provide mechanical power
to conventional air conditioning systems. Figure 4-19 shows a solar/
Rankine cycle used to power a conventional air conditioner (1, p.

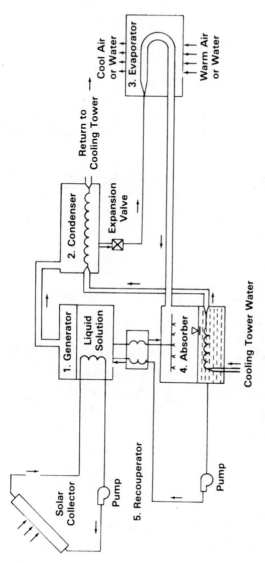

Figure 4-18. Absorption air conditioner. (From Ref. 9, p. 9-4.)

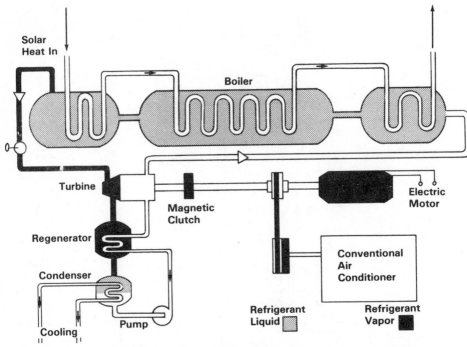

Figure 4-19. Solar/Rankine-powered air conditioner. (From Ref. 1, p. 162.)

162). Solar energy applied to the Rankine cycle at the boiler vaporizes a fluid which has a low boiling point; this vaporization supplies the power that enables a gas turbine to drive the shaft of an electric motor. A magnetic clutch allows the gas turbine to supplement the electric motor which is the source of power for a conventional air conditioner. As the vapor exits the turbine, it is cooled in a regenerator and passes to a condenser, where it is cooled and liquified by circulating water from a cooling tower (or by circulating air). The condensed low-pressure fluid is then pumped to a higher pressure, warmed in the regenerator and in the cooling jacket of the turbine, and reenters the cycle at the boiler. Because of the relatively low temperatures at the boiler, the solar/Rankine cycle is not very efficient; however, the energy that it provides, except for the parasitic energy required to operate the pump, is free.

SOLAR-ASSISTED HEAT PUMPS

A solar-assisted heat pump uses a higher temperature heat source
(typically, 40 to 110°F above the ambient air or water temperature)
to provide the heat source for a heat pump. In addition, because
the fluid temperature delivered from the collectors is usually low,
greater efficiency can be attained from the solar heat pump system
than from a solar system which uses direct heating methods and which
typically requires 150°F water temperature above ambient. Thus, a
solar heat pump system is appropriate in extremely cold, windy, or
cloudy areas, areas where flat-plate collectors could be used effectively to collect solar energy at temperatures sufficient for a heat
pump yet not effective for active heating systems.

A heat pump with a solar heating system can be installed according to a number of different arrangements (6, pp. 7-11 to 7-14).
The simplest would be to use a conventional heat pump as the auxiliary heat unit in a liquid system. This is shown in Fig. 4-20. This
arrangement includes the following operating modes:

1. In mild weather, the solar-heated duct coil provides heating.
2. As the temperature drops, for example to 45 to 50°F, the heat
    pump supplies heat and the solar unit goes into a heat-storage
    mode.
3. At the heat pump balance point (output of heat pump equals build-
    ing load) the duct coil is supplied heat from storage to supple-
    ment the heat pump output.
4. In the event that there is no solar heat at the collectors or in
    storage, electric resistance heaters are energized to supplement
    the output of the heat pump.

Another approach is to use a conventional water-to-air heat
pump, with solar assist. This is shown in Fig. 4-21. There are
three operating modes for this system:

1. If the storage water is 80°F or above, solar heating can be pro-
    vided through the duct coil.
2. If the storage water is between 80 and 45°F, the heat pump ex-
    tracts heat from storage, boosts the temperature via the refrig-

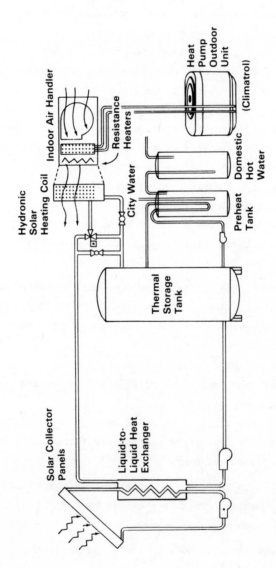

Figure 4-20. Air-to-air heat pump with solar assist. (From Ref. 6.)

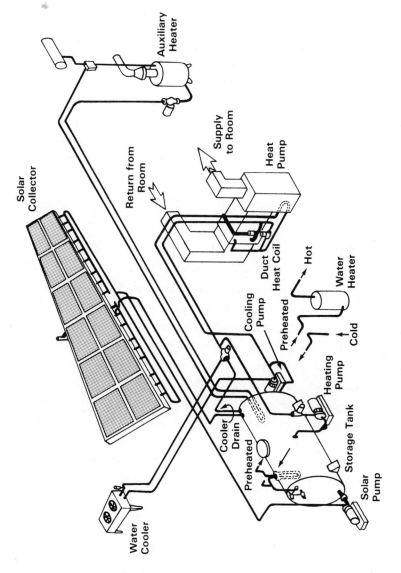

Figure 4-21. Water-to-air heat pump with solar assist. (From Ref. 6.)

eration cycle, and supplies heat through the indoor refrigerant
coil placed in the duct.

3. If the storage temperature falls below 45°F, auxiliary heat
    raises the temperature of the storage tank to 55°F.

The use of the water-to-air heat pump permits a greater drawdown of
storage temperature, which improves overall collector performance.

There is not yet a clear indication of which heat pump arrangement will prove best. In an air system, the heat pump can be most
advantageously used if it is operated simply as the auxiliary furnace, to raise the temperature of the air circulating from storage
to the rooms. As for the liquid system, whether the heat pump should
be used in a similar fashion or whether the heat source should be
the solar storage tank cannot yet be determined.

SOLAR SYSTEM SIZING

The first task in solar building design is to reduce the building's
overall energy requirements: The smaller the building's heat loss,
the smaller the demand on the solar energy system and the smaller
the necessary size of the collector area and storage volume. Unfortunately, this may result in a larger per-square-foot cost of the
system. "Active" solar energy systems, which use solar energy by
first collecting and then storing it for later use, are not yet
economically feasible for relatively small homes (1500 ft^2 or less)
and for buildings with relative small heat losses (15,000 Btu or
less per degree-day).

Although some designers are infatuated with indirect means of
collecting and using solar energy (such as the use of solar collectors and heat storage systems), natural and direct means of using
the sun's energy should be considered first. Though in some cases
they are more difficult to deal with in the design process, passive
solar techniques can be more effective, more comfortable, more efficient, and less expensive than active techniques.

After energy-conscious designs have been incorporated into a
house design, the solar heating system should be sized to provide

the desired fraction of the total heating load of the building. This
fraction can be chosen arbitrarily, or it can be determined from
economic analysis to minimize the annual heating cost of the solar
auxiliary system. The size of the collector area must be determined
and based on that, the storage size selected. The size of the auxiliary furnace is based on the design heating load and heat delivery
rate. The sizes of the pumps, blowers, and heat exchanges depend
primarily on the collector size and heat delivery rate.

There are various methods, from detailed computer programs to
rules of thumb, for determining the fraction of annual heating load
supplied by solar systems. Rules of thumb can be used as general
guidelines, after which manufacturers of components and solar systems can provide more detailed information and specific recommendations.

Rules of Thumb

Rules of thumb for sizing air and liquid solar systems are presented
in Table 4-3. The various components of the systems may be determined from the collector area sizes.

Fraction of the Annual Heating Load Carried by a Solar
Heating System

The collector size necessary to provide the desired fraction of the
annual heating load for air and liquid solar systems can be determined by using Fig. 4-22.

Let us suppose that we desire a solar liquid system to supply
80% of the annual load. From Fig. 4-22, the value of AS/L corresponding to f, the fraction of the annual heating load delivered by
the solar heating system of 0.8 is 0.9. Thus,

$$\frac{AS}{L} = 0.9$$

and the size of collectors needed is determined by:

$$A = \frac{0.9 \times L}{S}$$

Table 4-3. Rules of Thumb for Sizing

Solar Air Heating Systems	
Collector Slope	Latitude + 15°
Collector Air Flow Rate	1.5 to 2 cfm/ft² of Collector
Pebble-bed Storage Size	½ to 1 ft³ of Rock/ft² of Collector
Rock Depth	4 to 8 Feet in Air Flow Direction
Pebble Size	¾" to 1" Concrete Aggregate
Duct Insulation	1" Fiberglass Mininum
Pressure Drops:	
Pebble-bed	1.2 to 0.3" W.G.
Collector (12-14 ft Lengths)	0.2 to 0.3" W.G.
Collector (18-20 ft Lengths)	0.3 to 0.5" W.G.
Ductwork	~0.08" W.G./100' Duct Length
Solar Hydronic Heating/Cooling Systems	
Collector Slope	Latitude + 15°
Collector Flow Rate	~0.02 gpm/ft² of Collector
Water Storage Size	1.5 to 2.5 Gallons/ft² of Collector
Pressure Drop Across	0.5 to 10 psi/Collector Module
Collector	
Solar Domestic Hot Water Heating Systems	
Preheat Tank Size	1.5 to 2.0 Times DHW Auxiliary
	Tank Size

Source: Ref. 9, p. 14-2.

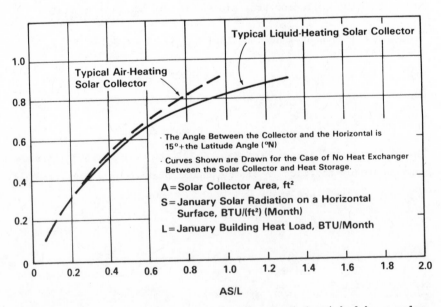

Figure 4-22. Fraction of annual heating load furnished by a solar heating system. (From Ref. 9.)

The desired size of the collector is dependent on the January solar
radiation S and January heating load L of the building. The mean
daily radiation for January is depicted in Fig. 4-23.

Sizing Example

   Problem: Determine the approximate collector size needed to
provide hot water for a family of four in a residential building in
Kansas City, Missouri.

   Solution: The average daily service hot water load in January
is:

$$L = 80 \text{ gal/day} \times 8.34 \text{ lb/gal} \times 1 \text{ Btu/lb/}^{\circ}\text{F}$$
$$\times (140^{\circ}\text{F} - 50^{\circ}\text{F}) = 60,048 \text{ Btu/day}$$

The desired service water temperature is 140°F, and the temperature
of the cold water from the main is 50°F. The total average solar
radiation S available in January, according to Fig. 4-23, is 680
Btu/ft^2/day. If the system is to provide 60% of the total annual
load, AS/L is about 0.5. Therefore:

$$A = 0.5 \times \frac{L}{S} = \frac{(0.5 \times 60,048)}{680} = 44.2 \text{ ft}^2$$

If 3 × 8 ft collector modules are available, 1.84 units would be required. Thus, two collector units should be used.

Sizing the Heat Storage Unit

The size of the heat storage unit is dependent on the material used
to store the heat and the size of the collector. Air systems generally require ½ to 1 ft^3 of pebbles per square foot of collector
area (6, p. 6-8), while water systems generally require 1 to 2 gal
of water per square foot of collector area.

   Example: A 260-ft^2 air collector array is installed in a house
with a 50,000 Btu/hr design heat loss and an overnight heat load of
287,140 Btu in January. Storage required: 130 (260 × 0.5) to 260
(260 × 1.0) ft^3 of rock.

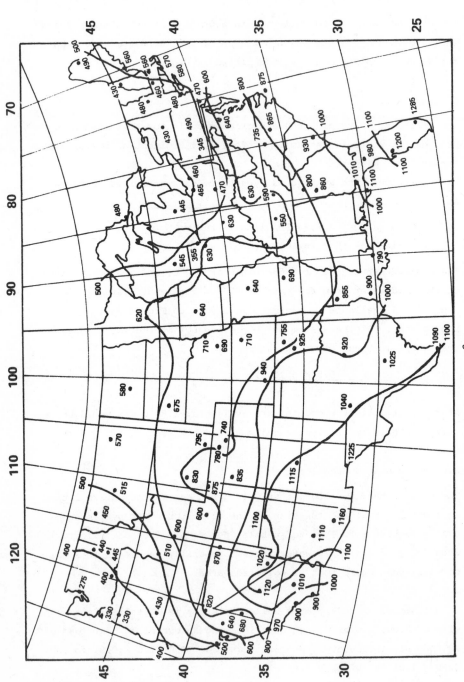

Figure 4-23. Average daily solar radiation (Btu/ft^2) for the month of January. (From Ref. 9, p. 7-28.)

At a specific heat of 20 Btu/ft^3 for rock and assuming a 70°F temperature rise, the storage would provide

130 × 20 × 70 = 182,000 Btu

or

260 × 20 × 70 = 264,000 Btu

Since the overnight heat load is 287,140 Btu, a 205-ft^3 pebble-bed storage bin, which falls within the ½ to 1 ft^3 guideline, would be adequate.

Sizing Heat Exchangers

Most manufacturers of heat exchangers use computer programs to select the properly sized unit, and selection is based on the flow and temperature conditions specified by the designer. To select a heat exchanger, the designer would have to determine:

1. The desired approach temperature (usually 10°F)
2. The collector loop flow rate (from collector manufacturer specifications, but usually about 0.02 gal/min per square foot of installed collector)
3. The amount of Btu's to be transferred (collector output at specified design radiation)
4. Either the minimum storage temperature or the fluid flow rate from storage through the heat exchanger (6, pp. 6-8 to 6-9)

Because of the interrelationships between collector, storage, and terminal unit (usually a duct coil), a complete analysis of system performance under actual conditions (laboratory or test house) must be made to determine the size of the heat exchanger. The designer/technician usually depends on assistance from the component manufacturers to properly select the best heat exchanger.

PERFORMANCE OF SOLAR HEATING AND COOLING SYSTEMS

It is rather difficult to determine who actually built the first
active solar-heated house in the United States: a number of claims
are made by solar activists. There are now thousands of solar energy
installations which run the gamut of heating, cooling, and hot water
applications. Many of these systems were built with private funding. However, a number of them were part of the Department of Energy
(DOE)/Housing and Urban Development National Solar Heating and Cooling of Buildings Program. Some of these systems are performing very
well, but a number of them are performing rather badly.

An excellent book describing 120 existing solar buildings has
been written by William Shurcliff. Shurcliff discusses in great detail solar heating systems which represent the best architectural
and engineering thinking in the United States (3). The buildings
chosen--houses, office buildings, schools--are representative of
buildings located in all parts of the United States--the East,
Southwest, Midwest, South, Southwest, and Far West--and seven of the
buildings are in Canada. A number of other books on solar applications are currently available to the public. Hence, this section
does not dwell on specific applications but concentrates instead on
the performance and reliability patterns of solar heating systems.

Stephen Sawyer, in his survey of 177 owners of solar systems,
attempted to determine their experiences, attitudes, and assessments
of the cost, performance, and reliability patterns of those systems
(10). Only individuals who satisfied seven specific criteria were
included in the survey. These criteria were adopted to ensure a
general uniformity in each consumer's system, involvement, experience, and financial commitment:

1. The consumer had to be the owner throughout the entire adoption
    process (purchase, installation, operation).
2. The consumer had to expect to remain in the house for the near
    future.
3. The system had to be "active," with a storage system for space
    and/or domestic water heating.

4. The residence had to be a single-family home.
5. The system had to have been purchased during or since 1973.
6. The system had to be entirely owner-financed, either directly or through a bank loan.
7. The installation had to be in either New England or the southwestern states of Colorado, Arizona, and New Mexico.

There is a tremendous diversity in the combined purchase-installation costs of the systems surveyed (see Table 4-4). The average cost of all types of systems is $4904, but there is high variability among the systems, as indicated by the very large standard deviation from the mean. Average system cost figures assume much more significance when disaggregated on the basis of system function. These average costs are $1476 for domestic water heating, $5279 for space heating, and $6669 for the combined systems. Once again, the figures are highly variable within each of these categories.

The average cost of the packaged systems exceeds the overall average in all categories. Conversely, the average cost of the homemade systems is less than half the overall average for each of the three functions. The cost variability of the site-constructed,

Table 4-4. *System Costs According to Function and Manufacture*

Function	Manufacture	Average	Standard Deviation	Number In Cell
Domestic Water	Overall	$1,476	$ 818	46
	Packaged	1,858	645	30
	Contractor	1,208	743	6
	Homemade	493	304	10
Space Heating	Overall	5,279	3,867	53
	Packaged	6,333	2,324	11
	Contractor	8,194	3,762	18
	Homemade	2,610	2,528	24
Combined	Overall	6,669	2,378	78
	Packaged	9,133	4,584	24
	Contractor	7,362	2,584	29
	Homemade	3,500	3,732	25

Source: Ref. 10.

contractor-made system is less definitive. It was anticipated that
the cost of these systems would fall somewhere between the costs of
the packaged and homemade systems, or be approximately 60 to 70% of
the packaged systems costs.

Another area in which the experience of solar consumers can
provide valuable information deals with the pattern of technical
malfunctions. On an aggregate level, 73% (n = 112) of the consumers
experienced some technical problems with cost of repair, which averaged $156, and with "downtime," which averaged about 12% of the
time. Many consumers indicated that their systems, while functioning adequately, were not operating at "peak performance levels" because of chronic technical problems. The general nature of the
technical problems is contained in Table 4-5. The malfunction experiences of the consumers indicate problems in over 50% of the systems during the initial 3 months of use and in over 40% of the systems after this period. The control systems were the most frequent
sources of problems.

The DOE, in its National Solar Heating and Cooling of Buildings
Program, has funded 475 single- or multi-family projects. About
one-third of these have been completed, most of which have been sold
or rented. These solar homes and systems are providing a variety of
information. Table 4-6 depicts the numbers and types of projects
included in the analysis by Freeborne, Mara, and Lent, of 23 of
these projects, the total of which represents about 7% of the grants
awarded (11).

Table 4-7 shows the apparent solar fractions, the amount of
solar energy provided by the system, for the four types of projects
included in the Freeborne, Mara, and Lent study. For many of the
active space heating systems, apparent solar fractions were below
those predicted in design. Solar fractions were generally lower
for air than for liquid systems, in which solar fractions ranged
from a high of 52% to a low of 12%, with an average of 33%. For
air systems, the range was between 45 and 4%, with a 17% average.

Table 4-5. Nature, Frequency, and Significance of Technical Malfunctions

Nature, Frequency, and Significance of Technical Malfunctions

	Frequency of Occurrence %	Frequency of Occurrence n	Problem Significance*
The Control System	41	63	1.9
Water Pumps or Air Fans	21	32	2.2
Leaks in Pipes to and from Collector	18	28	2.1
The Storage System	14	21	1.7
Collector Cover Glass	10	16	2.4
Leaks in Collector Seals	10	15	2.1
Heat Exchange System	9	14	2.1
Collector Cover Plastic	8	13	1.8
Fluid Leaks in Storage System	6	10	1.5
Insufficient Insulation	4	6	3.1
Corrosion of/in the Collector	3	4	3.5

Occurrence of Technical Malfunctions

	%	n
Number of Systems in Service	100	154
Number of Consumers with Problems	73	112
Within 3 Months of Installation	54	83
After Initial 3 Month Period	41	63

*Problem significance is calculated from the average value assigned by consumers experiencing the problem: 1 = most serious, 2 = second most serious, 3 = third most serious, 4 = all others. The lower the average number, the more significant the problem to those consumers

Source: Ref. 10.

Table 4-6. Numbers and Types of Projects

System Application	# of Projects
Space and Domestic Water Heating - Liquid	11
Space and Domestic Water Heating - Air	5
Domestic Hot Water	5
Passive Space Heating	2
Total	23

Source: Ref. 11.

Of the five domestic hot water systems studied, two had solar
fractions which were very high, while three reported solar contributions lower than expected. Both passive systems showed very high
solar fractions. Lower solar fractions in active space heating systems were usually correlated with significant design or installation
problems.

The ratios of measured to predicted auxiliary heating use with
solar energy are presented in Table 4-8. A few of the projects report substantial solar contributions but these are not quite at design levels. In spite of the fact that many active space heating
systems had apparent solar fractions that were lower than their
predictions, 6 of 11 liquid and 3 of 5 airspace heating systems used
amounts of auxiliary energy equal to or less than what was predicted.

Table 4-9 shows results for space heating projects in which
losses from storage and transport have been measured for the months
between September and April. Of nine projects, eight measured
losses of over 50% of the measured solar contribution, and three of
the eight lost more solar heat from storage and transport than was
delivered to the load through the regular distribution system. If
most of the heat loss from storage and transport had gone to the
conditioned space, the projects' solar fractions would have improved
noticeably. The solar fraction obtained by adding the measured
storage and transport losses to the apparent solar fraction is presented as the potential solar fraction in Table 4-9.

The Freeborne, Mara, and Lent analysis suggests that while measured solar fractions of many active space heating systems were below design specifications, uncontrolled losses in some projects helped to reduce heating loads and thus saved on auxiliary energy.

The dearth of knowledge about what is required in solar hardware has resulted in many unnecessary errors in the design, production, and installation of solar heating and cooling systems. These mistakes increase costs, reduce system efficiency and performance, or reduce system lifetime. Mistakes therefore extend the pay-back period, or can even prevent pay-back in the lifetime of the system. Only through elimination of these mistakes will the owner get the maximum benefits from a solar heating or heating and cooling system. Some means to ensure that the system is performing properly should also be provided, since improper performance of the system is one of the most difficult problems to detect.

CONCLUSION

The desirability of solar energy use in homes and buildings is obvious, but the road to these ends is not always smooth or easy. There are technical, legal, and economic difficulties, as well as homeowner attitudes that pose obstacles. Furthermore, there is a natural human reluctance to be "first in the water." Many homeowners and builders find themselves hesitant to invest in systems or try procedures with which they are unfamiliar. Fortunately, this will gradually change as more solar systems become operational and more hard data becomes available.

Designers have been deterred by lack of good input data on climate and on system performance; they have been concerned about the unavailability of reliable, warranted systems, as well as about the quality of the systems' installations. They have also been concerned about potential liability in case of failure. Builders and contractors lack experienced, qualified tradesmen and are afraid of cost overruns. There may be other, more profitable jobs available to them, and servicing may be a problem.

Table 4-7. Apparent Solar Fraction

Project	Period	Measured Solar Contribution (MMBTU)	Measured Solar Contribution & Measured Auxiliary Thermal (MMBTU)	Apparent Solar Fraction
A-1 (Perl-Mack)	Jan. 1978-Apr. 1979	20.41	51.38	40%
A-2 (Matt Cannon)	June-Oct. 1978, Dec. 1978	11.05	21.29	52%
A-3 (Saddle Hill)	Mar-Apr. 1979	7.75	19.39	40%
A-4 (Stewart-Teele Mitchell)	Dec. 1978-Apr. 1979	6.53	52.64	12%
A-5 (Ortiz & Reill)	Aug. 1978-Apr. 1979	8.09	16.58	49%
A-6 (Design Construction)	Feb. 1979-Apr. 1979	13.60	31.36	43%
A-7 (Sir Galahad)	Feb. 1978, May-June, 1978 Sep-Nov. 1978	2.73	8.45	32%
A-8 (Chester West)	May, 1978, July-Nov. 1978	3.96	9.74	41%
A-9 (Homes by Marilyn)	May, 1978, Sept-Dec. 1978 Mar-Apr, 1979	12.64	34.84	36%
A-10 (J.D. Evans - House A)	Oct-Dec. 1978, Mar-Apr. 1979	9.97	26.84	37%
A-11 (J.D. Evans - House B)	Oct-Nov. 1978, Feb-Apr. 1979	5.52	33.59	16%
	Total	102.25	306.1	33%

System*	Period			
B-1 (Frank Chapman)	Dec, 1978-Feb, 1979	1.76	31.66	6%
B-2 (Heliothermics - Lot 6)	Jan-Apr, 1978, Oct-Dec, 1978	25.71	56.64	45%
B-3 (Washington Natural Gas)	Nov, 1978-Apr, 1979	5.95	35.85	17%
B-4 (Moulder)	Dec, 1977-Feb, 1979	10.24	116.35	9%
B-5 (Suntech)	Oct-Dec, 1978	.95	24.61	4%
Total		44.61	265.11	17%
C-1 (Facilities Development)	Mar, 1978-Apr, 1979	113.31	474.47	24%
C-2 (A Frame)	Feb-Nov, 1978, Jan-Apr, 1979	8.74	9.61	91%
C-3 (Saddle Hill #77)	Nov, 1978-Mar, 1979	3.80	15.39	25%
C-4 (Saddle Hill #73)	Nov, 1978-Apr, 1979	3.33	9.45	35%
C-5 (Hel Wai Wong)	Apr-Dec, 1978, Feb-Apr, 1979	119.69	138.84	86%
Total		248.87	647.76	38%
D-1 (Colorado Sunworks)	Nov, 1978-Apr, 1979	37.38	49.37	76%
D-2 (Hullco)	Mar-June, 1978, Aug, 1978- Apr, 1979	78.18	78.25	100%
Total		115.56	127.62	91%

*System type: A = heating and water-liquid collectors; B = heating and water-air collectors; C = domestic hot water; D = passive.

Source: Ref. 19.

Table 4-8. Measured Auxiliary/Predicted Auxiliary Use with Solar

Measured Auxiliary Use (MMBTU)	Predicted Auxiliary Use with Solar (MMBTU)	Ratio of Measured/ Predicted Auxiliary Use with Solar
50.07	21.24	2.4
10.24	4.76	2.2
17.99	15.31	1.2
68.92	30.87	2.3
13.38	26.11	.5
2.05	18.24	.1
2.32	3.32	.7
2.59	3.29	.8
14.10	6.33	2.2
8.26	10.45	.8
12.48	5.31	2.4
202.40	145.23	1.4
29.90	80.18	.4
51.14	11.91	4.3
50.10	173.31	.3
106.11	41.41	2.6
12.47	10.81	1.2
249.72	317.62	.8
361.16	131.86	2.7
.87	7.38	.1
19.32	16.61	1.2
10.20	12.46	.8
21.33	9.20	2.3
412.88	177.51	2.3

No Predictions Available

Source: Ref. 11.

Uncertainties regarding the cost of the installed solar system and the subsequent return on capital investment have been major obstacles to marketing solar systems. The initial costs of solar energy systems are high compared to conventional systems, and zoning codes in some areas of the country prohibit exposed storage tanks and visible collector arrays. Some building codes place solar systems in the same category as steam boilers, putting insurance com-

System Type	Project	Solar Contribution (MMBTU)
Heating and Water Liquid Collectors	A-1	20.41
	A-2	11.05
	A-3	7.75
	A-4	6.53
	A-5	8.09
	A-6	13.6
	A-7	2.73
	A-8	3.96
	A-9	12.64
	A-10	9.97
	A-11	5.52
	Total	
Heating and Water-Air Collectors	B-1	1.76
	B-2	25.71
	B-3	5.95
	B-4	10.24
	B-5	.95
	Total	
Domestic Hot Water	C-1	113.31
	C-2	8.74
	C-3	3.80
	C-4	3.33
	C-5	119.69
	Total	
Passive Heating		

panies in the position of having to charge more to insure them than to insure conventional heating systems.

All of these current obstacles will someday be overcome and, as time progresses, the attractiveness of solar energy systems will be enhanced by the dwindling supply and corresponding increase of price of conventional fuels and by government programs and incentives. Conservation measures are being observed, and passive features are now being utilized. Active solar hot water systems, in general, and

Table 4-9. Profile of Measured Losses

Project	Period*	Measured Solar Contribution (MMBTUs)	Measured Losses (MMBTUs)	Measured/ Predicted Auxiliary Use with Solar	Apparent Solar Fraction	Potential Solar Fraction
A-1	Feb-Apr, 1978 Sept, 1978-Apr, 1979	8.71	32.9	1.3	33%	70%
A-3	Mar-Apr, 1979	7.73	5.45	1.2	40%	53%
A-4	Dec, 1978, Apr, 1979	6.53	4.03	2.2	12%	19%
A-5	Sept, 1978, Apr, 1979	6.79	4.44	.5	44%	57%
A-8	Sept-Nov, 1978	2.26	5.69	.5	43%	73%
A-10	Oct-Dec, 1978 Mar-Apr, 1979	9.97	2.33	.8	37%	42%
A-11	Nov, 1978	5.40	2.74	2.4	16%	22%
B-3	Dec, 1978, Apr, 1979	5.95	8.36	.3	17%	32%
B-4	Dec, 1977	5.49	5.10**	2.2	7%	13%

*Figures here are for the period listed only.

**Losses are probably much more extensive for this project but there has been difficulty measuring them.

Source: Ref. 11.

active solar heating systems, in some applications, have proven to
be cost-effective. Active solar heating and cooling systems will
be used more extensively as their economic benefits become more obvious. Components will be standardized and mass-produced, and suppliers will offer "packaged" systems. Designers, builders, installers, and service personnel will become more experienced.

Government programs will continue to support the growing solar
energy market. For example, Congress has established the mechanisms
and has provided funding for initiating and stimulating the national
solar energy program. To date, under the Solar Heating and Cooling
Demonstration Act of 1974, funds have been granted for the installation of a large number of solar energy systems for both residential
and commercial buildings. In addition, a solar tax credit of up to
$2200 has been approved.

With the continuing shortfalls in our energy supplies and with
the rapidly increasing cost of fossil fuels, incorporating energy
savings and solar energy features into all types of buildings is a
must for the future.

NOTES AND REFERENCES

1. *Solar Energy for Buildings Handbook,* University of Alabama,
    Huntsville, Ala., October 1979.

2. Bruce Anderson, *Solar Energy: Fundamentals in Building Design,*
    McGraw Hill, New York, 1977, p. 32.

3. William A. Shurcliff, *Solar Heated Buildings of North America:
    120 Outstanding Examples,* Brick House, Church Hill, N. H., 1978.

4. J. Douglas Balcomb, *State of the art in passive solar heating
    and cooling,* Proceedings of the Second National Passive Solar
    Conference, March 1978, pp. 5-12.

5. Deborah L. Buchanon, *A Review of the Economics of Selected
    Passive and Hybrid Systems,* SERI/TP161-144, Solar Energy Re-
    search Institute, Golden, Colo., January 1979.

6. Sheet Metal and Air Conditioning Contractors National Associa-
    tion, *Fundamentals of Solar Heating,* U.S. Department of Energy,
    August 1978.

7. Dennis O'Neal, Janet Carney, and Eric Hirst, *Regional Analysis
    of Residential Water Heating Options: Energy Use and Economics,*
    ORNL/CON 31, Oak Ridge, Tenn., October 1978, p. 11.

8. Solar Energy Applications Laboratory, *Solar Heating and Cooling of Residential Buildings: Design of Systems,* U.S. Department of Commerce, October 1977, pp. 19-20 to 19-21.

9. Solar Energy Applications Laboratory, *Solar Heating and Cooling of Residential Buildings: Sizing, Installation and Operation of Systems,* U.S. Department of Commerce, October 1977, pp. 14-3 to 14-5.

10. Stephen W. Sawyer, The cost, performance, and reliability patterns of solar heating systems: An assessment of 177 owners, Conference Proceedings--Solar Heating and Cooling Systems: Operational Results, Colorado Springs, Colo., Nov. 28 to Dec. 1, 1978, SERI/TP 49-063, Solar Energy Research Institute, Golden, Colo., 1979, pp. 325-329.

11. William E. Freeborne, Gerlad Mara, and Thomas Lent, The performance of solar energy systems in the residential solar demonstrative program, Preconference Proceedings--Solar Heating and Cooling Systems: Operational Results, Colorado Springs, Colo., Nov. 27 to 30, 1979, Solar Energy Research Institute, Golden, Colo., 1979, pp. 107-113.

5

Design, Our Visible National Resource

Charles R. Ince, Jr.
American Institute of Architects Research Corporation
Washington, D.C.

Imagine a building designed to be air-conditioned by natural means--
a vast glass and iron tent from which, within 2 min, stale and overheated air can be completely exchanged with fresh air, without benefit of artificial or forced ventilation. On the hottest August
afternoon the interior temperature rises no higher than 85°F.

That building exists today in Washington, D.C., but it was not
built by today's architects. It was designed nearly a century ago
by an Army engineer, General Montgomery C. Meigs, to house the U.S.
Pension Office (see Fig. 5-1). General Meigs' plans called for
stale air to be drawn out through ventilating lanterns in the roof.
All the windows were double-glazed, which helped retain heat within
the building during the winter. In summer, solar gain was reduced
by turning all roof glass to the vertical, in order to avoid the
greenhouse effect of sloping skylights. The building itself was
insulated by its masonry construction, construction that features a
series of space openings in the brickwork all around the building
under the windows.

The Pension Office was energy-efficient, at least until a series
of remodelings over the years resulted in the sealing of the transoms,

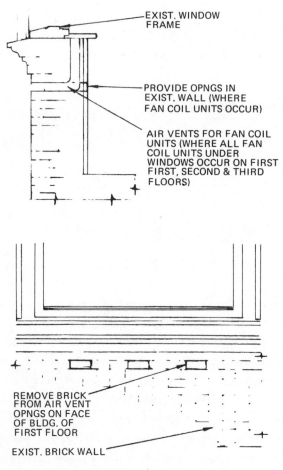

Figure 5-1. *Meigs' building plan detail.* (Courtesy of the National Building Museum.)

as well as the blocking of the ventilating holes in the brick. After that, it was necessary to put in an artificial air conditioner.

Here we are faced with the intriguing fact that a building constructed over a century ago is less energy-efficient today than when it was first designed. Nor is this a fluke. We know, for instance, that in New York City, buildings designed before World War II consumed half as much energy as comparable buildings designed after

the war. Even assuming that we, demanding that our working and living spaces be a constant 78°F in winter and 68°F in summer, are a
more pampered generation than our forebears, examples like the Meigs'
Pension Building suggest that we have gotten into a habit of designing our built environment in a manner that differs fundamentally
from past practice. A few examples illustrate the shape of this
difference.

Walk through Jefferson's Monticello. Notice the triple-sash
windows, the narrow interior stairwells, the underground utility
areas, the almost unbroken masonry walls on the north side complemented by the ample fenestration on the south side. Travel up the
East Coast to New England and take a look at that region's characteristic saltboxes. Again, one is struck by the inspired siting
and design (whether intuitive or planned) that make maximum use of
the natural energy which surrounds the building. The heating and
lighting of the sun and the cooling of the wind are important back-
ups to the man-made energy generated by stoves, fireplaces, and
candles. Indeed, it might be debatable whether it is the stove or
the sun that serves as the backup source of energy.

The prehistoric Pueblo structures in our desert Southwest complete this picture (see Fig. 5-2). These ruggedly handsome structures in their striking setting argue the view that a building, its
occupants, and their activities serve as integral elements in a co-
herent pattern of energy flow so natural that it seems to be coded
and filed somewhere in mankind's genes. In other words, wherever
we look--Virginia, New England, New Mexico--we discover that for
all their regional differences (and, as is pointed out later, *because*
of these very differences), these older buildings share one characteristic trait: By using the known technology of the time, energy
efficiency was designed for the purpose of comfort.

Today, as far as energy use is concerned, much of modern design
practice suggests a deviation from past concepts. Whatever else
the glass and precast iconography of Park Avenue or Main Street may
read, the message is seldom energy-conscious design. What prompts

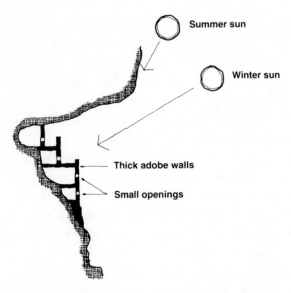

*Figure 5-2. Pueblo cliff dwelling. (From Solar Dwelling Design
Concepts, U.S. Department of Housing and Urban Development, GPO#
023-000-00334-1.)*

this collective amnesia among so many of our designers, builders, an
and their clients?   There is, of course, no single answer, but certainly the short, heady reign of cheap energy figures importantly.

For most of recorded history, energy (both man-organized and
man-made) was a precious resource.   The availability of a reasonably reliable bank of energy confined much of our most creative activity to a temperate zone.   There, human needs could achieve, however painfully at times, a balance between the vicissitudes of climate and the sources of energy, such as wood and charcoal.   Whenever
this delicate balance was dramatically upset, as in the case of pro-
longed drought or cool periods, civilizations disintegrated as the
people migrated in the direction of more abundant sources of energy.

The call of man-made sources of energy was even more precarious.
Only the very rich could afford to squander the limited supply of
fuel, and then usually only at the expense of a much larger group
whose reward was famine and disease.   To cite an unhistorical but,

in this case, useful analogy, Nero had time to pursue the fiddle because he was willing to have his Romans pay an enormous price.

The Industrial Revolution appeared to change the course of history in this regard. Technology unlocked what seemed at the time
to be a vast, almost unlimited store of man-made energy. The impact
of this revolution became especially pronounced after World War II,
for the succeeding years saw the development of mechanical and electrical systems that relied on cheap, nonrenewable sources of energy.
No longer constrained by site or climate, designers could turn away
from their drawing boards with plans for structures that could, with
little or no modification to the design, be built in Boston, St.
Louis, or San Diego. If necessary, a large furnace could be squeezed
in or more cooling units added to provide an artificial environment.
Whatever nature might unleash, there was always the thermostat. Of
course, not all designers went this route. Not every window was
sealed. Nevertheless, like Meigs' Pension Building, too many were.

Many of us were born about this time, a period which saw the
cost of energy steadily decline. But 1970, the year that opened a
new decade, signaled the dawn of a new, disturbingly different era.
A decline in energy supply, an increase in demand--this coupled with
the Arab oil embargo and the resulting long lines at gas stations
in the winter of 1973-1974-- forced many of us to take a new, energy-
conscious look at our life-style. And since buildings, or, more accurately, since the people within buildings use 25 to 33% of the
energy that is consumed in this country, it is not surprising that
the making and the running of our built environment came under close
scrutiny. Figure 5-3 provides a summary of energy usage by consuming
sector.

It is also not surprising that our initial response was conditioned by our immediate past. Technology had brought us cheap energy; technology would bail us out of this difficulty. In other
words, just as it had been after World War II, the supply of energy
was once again not considered to be a design issue.

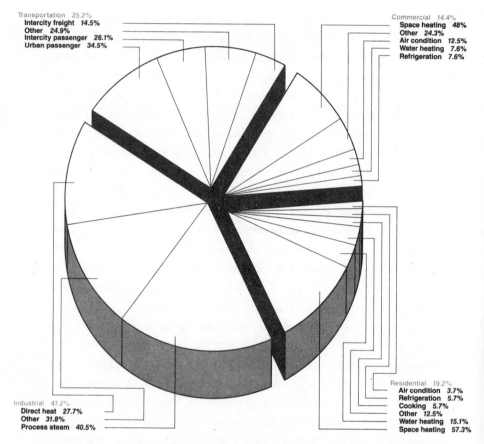

Figure 5-3. Energy consumption patterns. (Courtesy of Fred Greenberg, Art Director, AIA Research Corporation, Washington, D.C.)

One school of thought believed that the solution to the problem lay in mechanical systems (1). Photovoltaic cells, solar furnaces, trickle-down collectors—these were considered to be the wave of the future. Based on this assumption, designers felt that there needed to be only a temporary concern with rising fuel costs and the bleeding of dollars out of the country for foreign oil.

Another approach, again technological, focused on tightening the structure of buildings. Insulation, weather stripping, airlock entries, double-glazing, and a whole arsenal of prophylactics would

keep the north wind at bay and conserve our limited store of fossil
fuels. These measures would also cut down on the troublesome rate
of increase of monthly heating and utility bills.

In neither of these fundamentally mechanical approaches was
there much of a role for design. But as the federal government soon
discovered, design was not a resource that could be ignored. In
the early cycles of the Department of Housing and Urban Development/
Department of Energy (HUD/DOE) solar heating and cooling demonstration programs, a number of proposals for active energy system projects were made, many of which results in collectors being constructed atop conventional buildings. As these projects reach completion
and were put into effect, researchers discovered that although the
active systems might function properly (which was not always the
case), the buildings themselves were often energy-inefficient.
Some factor in the equation that should have led to energy efficiency was missing.

What was missing were the lessons to be learned, not from the
immediate past, but from farther back--at Monticello, in New England,
in New Mexico, in the Pension Building of General Meigs. Here were,
literally, sermons in stone preaching that design had, in fact, a
role, an important role, to play in the new era we had entered. As
I would like to argue in the rest of this chapter, this era need not
be feared as the dawning of austerity, sacrifice, a shrinking of
the quality of life. Instead, now that that brief aberrance in our
history--those decades of cheap man-made energy--is behind us, we
face an important opportunity. This opportunity holds out the
promise of at least a 40% reduction in the energy consumed by buildings by applying the principle of what we already know or once knew.
But more than that, it is an opportunity to recall a sense of place,
to enhance the quality of life, and to usher in a new design which
is potentially more far-reaching and revolutionary than any since
the Renaissance. We must rediscover the fact that design is one of
our most important national and natural resources.

DEFINITION OF TERMS

To talk about energy as a design issue means to rethink the framework of the energy problem as it has been discussed so far. In the
first place, we need to stop thinking about energy consumption in
the narrow sense of fossil fuels. In assessing energy from a broader
point of view, which is the harnessing of forces to do work, one
realizes that there are many other energy sources, such as the sun,
the wind, water, and even, the economic energy generated by capital.

A more difficult step to take in reframing the current discussion of energy, a step that brings us to the heart of the design issue, is the substitution in our lexicon of *energy consciousness* for
energy conservation. What is the difference?

The route of energy *conservation* is basically a mechanical approach. On the other hand, energy *consciousness,* enlists the unique
talents of the designer. Unlike energy conservation, which implies
sacrifice and cutback and, in terms of the design process, implies
that one can design as usual and then modify as necessary, energy
consciousness describes a value system intrinsic to the whole design
process and relates to the desire to have buildings provide comfort
compatible with the environment. Energy-conscious design requires
that buildings interact dynamically with their surroundings, enhancing and controlling them as comfort requirements indicate. This is
the dynamic relationship that animates the design of General Meigs'
Pension Building in Washington.

Whereas energy conservation is identified with systems efficiency or a *quantitative* approach, energy-conscious design implies a
qualitative approach to building design that includes:

1. Integration of energy into initial design decisions concerning
    program, location, and form
2. Reliance on natural and renewable resources to supply energy
3. Maximum efficiency in system design and integration

Such an approach has as its goal structures that make maximum use of
the natural, *free* energy already present in the environment. The

idea is to design buildings that require few mechanical parts or complex hardware to maintain a comfortable climate for people inside. This is what is meant by the term *passive design* (2).

Passive design takes its cue from the example of nature. In his book *Energy and Form,* University of California researcher Ralph Knowles draws upon the plant and animal world to point out that shape or form is the response of an organism to energy, a response that is in turn modified by its proximity or location to the source of energy as well as by its own metabolism (3). Applying this organic metaphor to building design, Knowles develops a holistic approach to energy-conserving design that has as its key elements *location* (orientation and siting), *form* (shape and structure), and *metabolism* (mechanical and electrical systems) (see Fig. 5-4). It is not surprising then, that it is precisely a structure's relationship to its environment that distinguishes most sharply the difference between energy conservation and energy consciousness (4). For example, a tenet of energy conservation is to seal a building which might be capable of relating and responding to its environment, a building which could use the sun's energy for light and heat, the wind for ventilation and cooling, humidity for its impact on comfort, and temperature for the energy inherent in its day-night diurnal swings.

But, if a strict conservation program has its limitations, there was also a point not too long ago when advocates of passive design were similarly narrow in their outlook. Perhaps in reaction to the early parochialism of active system research, passive designers at first defined a passive energy system as one capable of controlling the flow of thermal energy into, through, and out of a building solely by natural means, and they followed that definition to the letter. They avoided all mechanical elements in their energy systems and, like their active counterparts, came up with mixed results. Their work did produce some innovative and refined design concepts. For instance, they carried the principles and jargon of passive design—direct gain, sunspace, siting, orientation, trombe walls,

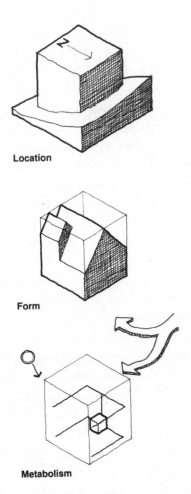

Location

Form

Metabolism

Figure 5-4. Energy and form. (Courtesy of Ralph Knowles, University of Southern California.)

thermal massing—into the practitioner's lexicon. Also, for the first time passive projects from coast to coast defined the regional and climatic guidelines of energy-conscious design (5).

For example, in the fall of 1976, the American Institute of Architects (AIA) Research Corporation sponsored and administered a solar dwelling design competition, funded by the Exxon Corporation, for architectural students in the United States and Canada. Of the more than 2000 students who participated in the competition, 10 were finally selected by an interdisciplinary jury to receive awards for their designs. All of the winning designs were innovative, but two in particular deserve special attention. Submitted by a single entrant, both are the work of Mike Marsh, a Canadian architectural student at Nova Scotia Technical College in Halifax.

Marsh's two passive solar dwellings, designed for two climatic extremes, bear little resemblance to each other. Nor should they. Each is designed in response to its particular climate--one to keep warm in cold weather, the other to keep cool in warm weather. The two sites are Nova Scotia and Jamaica.

The plan for the house in Nova Scotia is shown in Fig. 5-5. The design takes advantage of the earth's insulating qualities in helping to reduce temperature fluctuation in the house. Setting the north side of the building into the slope of a hill cuts heat loss by reducing the amount of exposed area. Sod covers the north slope of the roof.

The site is used to advantage with coniferous trees located to the north to block the winter winds. Deciduous trees to the south shade the house in summer, but allow the sun to warm the house when they lose their leaves in winter.

Large glazed areas on the south serve as passive solar collectors. The sun radiates through the glass onto a large, dark-colored thermal (mass Trombe) wall. The wall stores the heat, then slowly reradiates it later into the living space. Heated air between the Trombe wall and the glass heats the large living space by convection.

The windows are triple-glazed with thermal drapery to reduce nocturnal heat loss, and there are few operable windows because cold infiltration is a major contributor to heating load.

A Heatilator fireplace provides the house with auxiliary heat. Heat radiates directly from the fireplace, while air ducts through the chimney capture heat that usually escapes through the flue. The house also has a supplementary electric baseboard heating system which derives some power from a wind-driven generator.

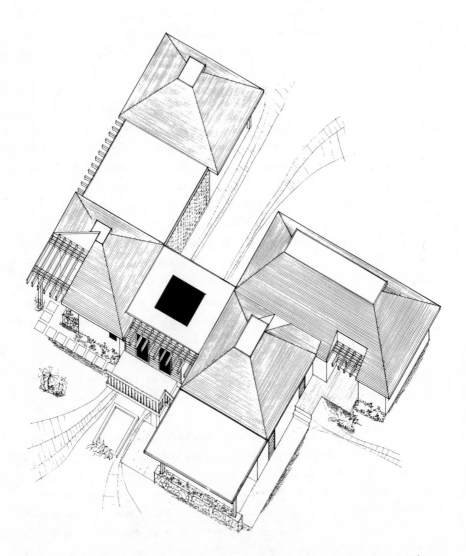

Figure 5-5. Marsh passive solar house--Jamaica. (From Capturing
the Sun, AIA Research Corporation, Washington, D.C., 1977. Courtesy
of Mark Marsh, Ontario, Canada.)

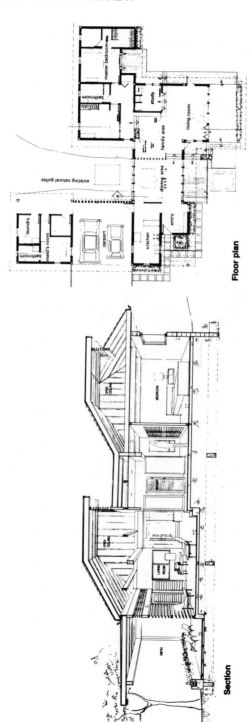

Floor plan

Section

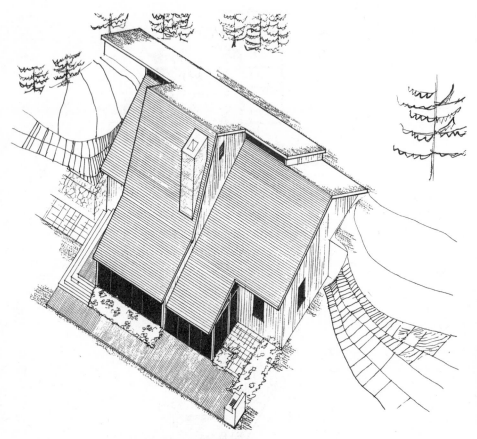

Figure 5-6. Marsh solar dwelling--Nova Scotia. (From Capturing the Sun, AIA Research Corporation, Washington, D.C., 1977. Courtesy of Mark Marsh, Ontario, Canada.)

Marsh's Jamaica house, Fig. 5-6, deals with an entirely different climate. It is concerned not with heating, but with cooling
loads and ventilation, so Marsh sought to control the effects of
solar radiation with louvers in glazed areas, overhangs, and high
ceilings that will allow heat to rise. Openings are located to
catch prevailing southern breezes and so provide an evaporative
cooling effect inside.

The large overhangs are complemented by careful landscaping to
shade walls and the surrounding ground from the sun. A self-shading

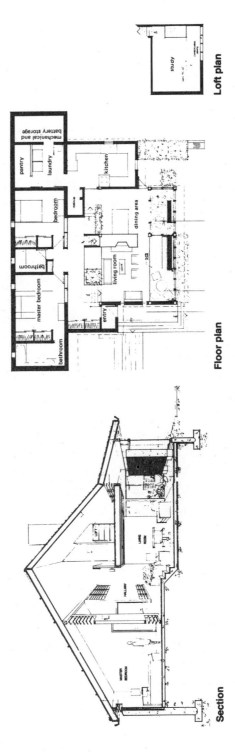

Loft plan

Floor plan

Section

concrete block is used similarly and adds thermal mass to the building, thereby minimizing interior temperature fluctuation.

Flat-plate solar collectors linked to a cistern for rainwater
are used to supply domestic hot water.

These two approaches provide us with a clear message on passive
design but, on the negative side, comes the realization that the
low-temperature limitations of pure passive design make it more appropriate for residential and small-scale commercial applications.
Furthermore, passive design alone often fails to maximize energy
efficiency. However, with mechanical assistance--not in the collection of energy but, for example, in the transfer of solar-heated air
to cold parts of a building--the performance of passive design could
soar. Energy conservation research, active technology research,
and passive design research developed their individual techniques
separately, and what has hampered our ability to design buildings
that are energy-efficient in the past was our failure to crossbreed
the three approaches.

Today, that crossbreeding is taking place. Designers are developing hybrid energy systems which combine conservation techniques
and active mechanical elements within a passive framework--a synthesis these designers think of as architecture's true leading edge.
This synthesis of "soft" design and "hard" technology is one that
architects and engineers have long sought. It represents the merg-
ing of three disparate streams of modern research with the culmina-
tion of that research which had as its model the traditional architectural goals and indigenous design solutions that date back to
the pueblos of New Mexico and the saltboxes of New England.

The federal government helped stimulate this crossbreeding with
the passage, in 1976, of the landmark Energy Conservation and Production Act, Public Law 94-385. This legislation, specifically
Title II, which mandates the reduction of energy consumption in
built environments, triggered what has been called the largest project of architectural research in U.S. history. The object of this
project is to lay the groundwork for standards that will affect the
energy performance--and likely the shape--of every building erected
in America from 1980 on.

ENERGY PERFORMANCE STANDARDS

Of particular interest to designers was the caveat in Public Law
94-385 that the proposed energy standards be performance-oriented
rather than prescriptive (6), and that the standards focus on whole
buildings rather than on their parts, as is usually the case in
present-day codes.

The following current analogy illustrates the differences between performance and prescriptive approaches: The Department of
Transportation (DOT) sets fuel consumption targets for new cars
which automotive designers are required to meet, according to a
schedule worked out between DOT and the manufacturers. DOT does not
dictate *how* car manufacturers should design their products--only
that those products achieve specified levels of fuel efficiency.
The approach--a performance approach, does not restrict the art of
automobile design; it accomodates and encourages the development of
technology and design innovation.

During the legislative debate prior to passage of the Energy
Conservation Act, it was successfully argued that performance standards, unlike prescriptive standards, would neither constrain design
nor freeze technology at a certain limited point. It was also argued that the performance approach would do more to encourage conservation. Thus, Title III of Public Law 94-385 called for design
standards in terms of goals to be met, without specification of
methods, materials, or process. These standards would encourage
both maximum practicable improvements in energy efficiency and increased use of nondepletable energy resources. Specifically, the
standards would be applied at the design stage of each building,
before its construction; they would take the form of energy budgets,
putting a ceiling, for example, on consumption of Btu's per square
foot per year; they would reflect the relative difficulties of conserving energy in different climates; and they would be based on the
intended uses of building types.

In order to develop standards that satisfied the intent of the
legislation, DOE and HUD had first to define "performance." A new
term was coined: *designed energy performance*--the energy consumption

of a building *as estimated* from the design, using a typical set of
assumptions about building occupancy and operation, as well as
weather and climate. HUD and DOE also had to determine the designed
energy performance of buildings in three situations: (1) as originally designed in 1975-1976, after the oil embargo; (2) as hypothetically designed according to component energy standards existing in
1978; and (3) also hypothetically, as designed to the maximum levels
of energy conservation practicable by designers today. With quantification of the three situations in hand, HUD and DOE could develop
interim energy performance standards that could be tested.

What was needed to do this was information, information that
did not exist. To develop this data, in 1977 the DOE contracted
with the AIA Research Corporation to conduct the necessary research
and provide the technical information needed to support HUD and DOE
policy decisions.

The first phase involved the collection of energy use data on
1700 buildings which were designed and being constructed in 1975-
1976. These buildings were taken from a statistical random sampling
representing a cross-section of building types and climatic regions
and comprised the first generation of buildings designed after the
1973 oil embargo for which complete data were available (the assumption being that energy conservation became of increased importance
to both designers and their clients in the shadow of the embargo and
rising costs). The data were then used for computer stimulation of
the buildings' designed energy performance.

In order to measure the energy performance (heating, cooling,
and lighting) of these buildings, a full range of data on construction characteristics and use requirements was carefully collected
from commercial building designers via a survey form. Information
on residential construction was collected from home builders and
mobile home manufacturers. The computer simulations that resulted
from this analysis helped to graphically illustrate the range of
energy performance for each building category. For the first time
there was documented evidence on the actual relationship or architectural design to energy consumption for the years 1975-1976.

What remained were questions of how the same buildings would use energy if they were designed in accordance with energy component standards in existence in 1978, and how the buildings would use energy if they were designed to the maximum levels of energy conservation obtainable from designers at that time. These were questions for which new research procedures had to be developed, procedures capable of generating and correlating information on *hypothetical* designs. This was Phase 2.

From the original 1700 buildings, 168 were selected for a second study. The designers of these 168 buildings, the architects and engineers, contracted to redesign these buildings in an effort to obtain maximum technically feasible levels of energy conservation based on the original building design and construction budget (adjusted 1978 dollars). When there were questions about modifying the original budget, the teams were instructed to use their professional judgment to determine whether increased costs could be justified to their clients.

The AIA Research Corporation acted, in a sense, as surrogate client. If the researchers could be convinced by the designers that a given change in the original conditions of the project were justified in terms of energy savings, then that change was permitted. In short, in Phase 2 these architects and engineers were asked to rethink the design of their original buildings, this time with the added requirement that they be conscious of the energy implications of each design decision. The emphasis was clearly on total redesign rather than on mere retrofit.

The objectives of the redesign program are perhaps best conveyed in the rules established for participants:

> We are asking architects and engineers to add energy conservation to their normal design criteria and to emphasize it in their approach to building design. We are not asking for an academic solution that emphasizes energy at the cost of human comfort and performance. We are going to designers instead of to a laboratory because we want good building designs that are also energy conscious. (7)

Along with their redesigns, the designers were also asked to provide detailed information, so that energy performance could be simulated and compared to that of the original buildings. Through this comparison, the maximum levels of energy conservation that are possible today could be determined.

By rethinking the design solutions, by allowing *how* a building's occupants used energy to dictate form, we learned that it was possible for architects and engineers to reduce the designed energy requirements by an average of 38% (8). In many cases, the cost of the building was reduced and, in almost every case, the general design character of the building improved.

The favorable results of the energy-conscious design process did not occur because the designers involved in the redesign added energy-conserving features to their buildings. Rather, they came about because the designers viewed the requirement to conserve energy as an opportunity to better understand and to explore a buildings' relationship to its environment. The interaction between a building and its environment and the manner in which the building's occupants extend their space into that environment were concerns that guided most of the conceptual development of the redesigns. Clearly, these concerns are also relevant to issues of urban redevelopment, surburban sprawl, and preservation of the natural environment. The redesign process revived an awareness of traditional, fundamental architectural design methodologies which can be used to confront major problems that influence the way people live.

The redesign process also helped to change the way in which the participating architects and engineers perceived a building and its development. Frequently, the design analyses included hour-by-hour sun, shade, or wind studies that helped the architects generate form in response to changes over time and space. These studies acknowledge the concept that a building is not a static entity and should not always be represented as such, but that it changes visually and functionally with cycles of environmental stress and seeks to act as the balance between them. Here again, the organic metaphor is applicable, because the structure is in this way responding like a living thing. And since these cylces of environmental stress

are predictable and measurable, designers can anticipate them and
use them to advantage.

The research project that had been initiated to gather data for
HUD and DOE also revealed that the characteristics of energy flow
are not only related to patterns of building form and structure, but
to the functional patterns that characterize energy use. Controlling use patterns in order to conserve energy not only affects a
building design; more fundamentally, it affects the program requirements that initiate that design. Here then emerged an opportunity
not only to improve a building's energy performance but its performance as an educational system, a home, a place to work, and so on.

This far-reaching exercise in creative research, in which data
and methodology were developed simultaneously, established the
theoretical credibility of energy as a design issue. But how does
actual practice, the *application* of theory, compare to the optimistic picture sketched by computer simulation? Rather well, it turns
out.

THE PRACTICE OF DESIGN: CASE STUDIES (9)

Residential

Already, many homes have windows to the sun, although few homeowners
realize how useful this sunshine can be in providing free and natural heating. The living room window, for example, if placed in
the best of all positions, facing the south, can provide as much as
100% of heating requirements. However, as soon as the sun goes down,
that solar heat disappears, unless the sun's heat is collected and
stored so that it can last through the evening. This can be accomplished by using direct-gain passive solar systems, which use dense
or heavy materials such as stone, concrete, or clay tile in the
building's walls, floors, or ceiling interiors to store some of the
sun's heat. An efficient direct-gain heating system also uses
shutters or insulating curtains on the windows.

The use of the living room window, in combination with heavier
floors, walls, or ceilings, and movable insulation or heat loss

control is the first type of passive solar heating system illustrated
in the following case studies. Moving the collector and storage
area to the face of the building for exterior wall or roof storage,
or completely away from the building in the form of greenhouses or
collector modules, determines a second and third type of passive
solar heating system.

Two other passive heating systems are in use in this country.
For time-lag heating, thermal mass used for exterior walls or roofs
can delay daytime heat gain in climates with significant day-night
temperature swings. In other climates, buildings underground or
earth berming can take advantage of relatively stable earth temperatures to provide annual time-lag heating. Each of these systems
involves its own set of controls against heat loss and overheating,
controls both to slow down or to speed up heat flow as well as controls to maximize heat gain.

Residential Case Study 1. A Guilford, Connecticut residence designed
by Paul Lytle, David Conger, and Steven Conger of Leela Design, Inc.,
satisfies 40% of its heating requirements through a direct-gain passive solar system worked into an intriguing overall design (see Fig.
5-7).

The crux of the system is the 35-ft bay window-wall, double-
glazed, on the building's south side. At the first floor level,
sunlight streams through the bay onto a radiant concrete slab above
a 21-yd^3 rock thermal storage area. Heat stored there radiates upward through the 2800-ft^2 home's three floors and is eventually
caught at the upper level and recycled down to the rock storage bed
for reheating. The recycling is accomplished with a thermostatically
controlled fan that modulates heat flow through the house. A window-
less, full-height air column on the north side buffers the house
from inclement climatic elements.

The house has both oil furnace and wood stove for auxiliary
winter heating. In summer--the site overlooks Long Island Sound--
all glazing areas open to transmit breezes, and night air blown
through the rock storage provides additional daytime cooling.

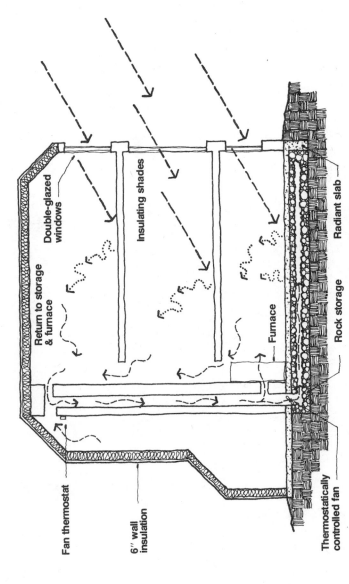

Figure 5-7. Leela design house. (From Ref. 1; designed and built by Steve Conger, Dave Conger, and Paul Bierman-Lytle, Guilford, Conn.)

Residential Case Study 2. The Kelbaugh house in Princeton, New
Jersey, designed by Doug Kelbaugh, features an indirect-gain passive
solar system (see Fig. 5-8). In such structures, the fabric of the
house continues to collect and store solar energy, but the sun's rays
do not travel through the living space to reach the storage mass.
This eliminates the direct gain limitation in which solar collection
temperatures are limited by occupant comfort needs, but it introduces

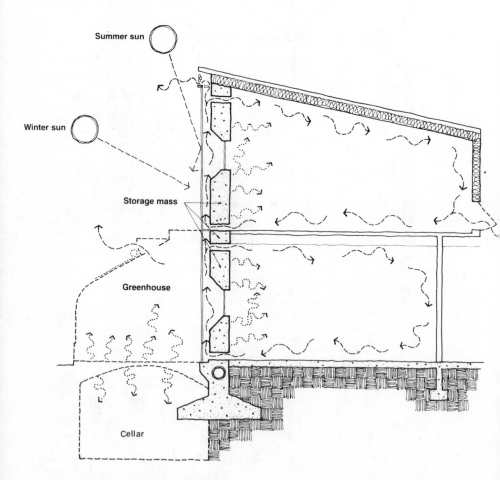

*Figure 5-8. Kelbaugh house. (From Ref. 5. Courtesy of Doug
Kelbaugh, Princeton, N. J.)*

the element of time lag. In the indirect-gain passive solar building, the storage mass collects and stores heat from the sun and
transfers heat to the living space in a design-controlled fashion.
Kelbaugh's 2100-ft^2 three-bedroom residence uses massive Trombe wall
construction inside a large glass facade in order to provide storage
for passive solar heat gain, and includes windows for south views
from each room (10). The all-glass south facade and the small solar
greenhouse are double-glazed, while all other building faces have
minimum glass exposure. The vertical 15-in. thick concrete wall
with a selective black coating absorbs the radiation entering the
600-ft^2 of glass on the vertical south face. In each room the air
is heated as it passes through the 6-in. space between the glazing
and the wall; the air then reenters the room through slots at the
ceiling level, providing natural solar heating for the rooms. Long-
term heat distribution is also provided by this massive storage
wall which, after many hours' delay, radiates heat to the room for
nighttime use.

   In summer, the convective distribution potential of this mass
Trombe building allows for induced ventilation. Vents to the outside at the top of the Trombe wall and in the greenhouse allow solar-
heated air to escape, drawing cooler air into the house for natural
ventilation.

   Deciduous trees also provide summer shading and complementary
elegance to this passive solar home. Because of the simplicity of
its design and orientation and because of its ease of maintenance
and operation, this house was able to reduce heating costs by 76%
the first winter and by 84% the second (10, p. 57).

Residential Case Study 3. A second indirect-gain building type is
designed around a roof pond. Here, the passive collector and storage mass have been relocated from the floor and the wall of the
building into the roof for radiant heat distribution to the living
space. The roof pond solar heating system is especially effective
in climates that have an overheated winter sun and less snow cover,
that is, in lower latitudes.

A cluster of small cabins of 285 ft^2 each was constructed in
Sonoma County, California, in order to study, on a comparative basis,
different generic forms of passive-solar space heating systems (see
Fig. 5-9). Designed by Peter Calthorpe of the Farallones Institute,
the cabins were built using the standard wood frame construction of
California and were well insulated to R-11 in the walls and to R-19
in the ceiling (11). One of these cabins demonstrates the effectiveness of the roof pond. This Sky Therm system, invented by Harold
Hay, uses 9 in. of water mass encapsulated in a vinyl bag custom-
made to horizontally fit a 13 × 4 5-in. area on the roof. Under the
bag is a ½-in. ferrocement ceiling of high thermal conductivity.
The sides and a lid for the bag are packed with 3 in. of urethane

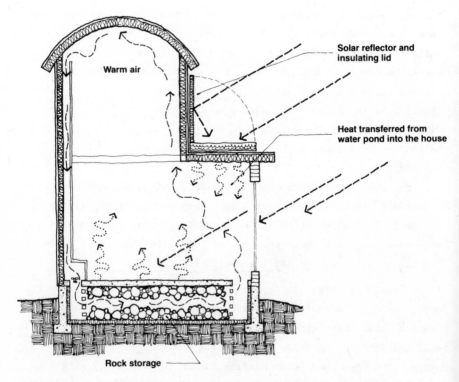

*Figure 5-9. Calthorpe cabin. (From Passive Solar Heating, U.S.
Department of Energy; Peter Calthorpe, architect, Inverness, Calif.)*

insulation. During the winter, the hinged lid is lifted to expose
the water to sunlight. A reflective surface directs additional
radiation onto the water. The lid is closed at night to control
heat loss, and the bag in turn radiates its heat directly to the
space below through the thin ceiling, which has negligible thermal
resistance. On the other hand, during the summer, the lid is closed
to protect the water from sunlight but opened at night, so that the
water pond can radiate its heat outward to the cool night sky, offering cool temperatures throughout the summer day.

Although temperatures vary between 64 and 78°F, the cabin remains comfortable throughout most of the winter and all of the summer. This roof pond building provides both passive solar heating
potential in winter and night sky radiant cooling potential in summer, leaving its own south facade free to provide a clear view of
the Northern Pacific forests.

Residential Case Study 4. New construction is not the only bene-
ficiary of an approach that enlists the services of the designer in
confronting the question of energy use. Virginia contractor Tom
Rust and his wife, Susan, had to conform to district regulations in
renovating their home in Alexandria's historic Old Town district.
Their solution: A narrow, brick street facade that is in keeping
with its colonial neighbors and which harbors a three-story central
atrium designed to heat and light the house (see Fig. 5-10).

In winter, the atrium gains solar heat directly and warms to
80°F, at which point automatic dampers open to duct the warm air
throughout the 3000-ft^2 house. In summer, the atrium exhausts hot
air up and out of the house. Between the atrium's contribution and
an active water-type solar collector on the roof (connected to a
1000-gal storage tank), up to 60% of the Rusts' heating requirements
are met by the sun.

Commercial

Commercial applications of passive technology may be one of the designer's most exciting challenges. But beyond zoning for different
consumption levels at building core and perimeter, weatherizing the

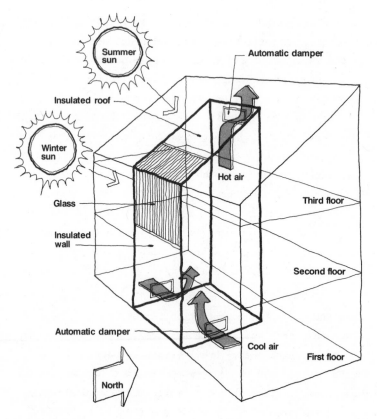

Figure 5-10. Rust house. (From Ref. 1. Courtesy of Tom Rust, solar design builder-owner, Alexandria, Va.)

perimeter against the external environment, and dabbling in active solar systems which lend themselves to the sealed building concept of conservation, designers have pretty much left most commercial energy considerations to the engineers. Not without reason. Passive energy technology is inherently low-temperature and small-scale, instinctively appropriate for residential application. Engineered systems in general are more at home in commercial construction, in which the installation and maintenance problems that can plague a homeowner are handled by a knowledgeable building engineer.

Nonetheless, as the research project conducted by the AIA Research Corporation for HUD and DOE demonstrates, design can significantly reduce the energy consumed by our hospitals, schools, and office buildings, at the same time generating an architecture that conveys a sense of place because it is finely tuned to the unique characteristics of its location. Already we have demonstrations of both possibilities.

Commercial Case Study 1. Recently, when the Georgia Power Company outgrew its original hearquarters, it decided to consolidate its corporate operations in a new central office building in downtown Atlanta (12) (see Fig. 5-11). The company turned to the firm of Heery and Heery Architects and Engineers, Inc. to develop a structure that would be a model of energy-conscious design. What this enlightened client received for its challenge was a building that is designed to meet the needs of human comfort by using only 55% of the energy required for comparable buildings designed in the recent past. Moreover, computer analysis revealed that this savings would be achieved with little or no effect on first cost.

In their design, the architects inhibit solar heat gain by reducing the use of transparent glass to a functional and psychologically desirable minimum. On the east and west facades windows are eliminated altogether, and on the south side they are shaded in the summer by setbacks and sunshade tubes. What the designers have recognized is that no two sides of a building respond to the environment in quite the same way.

The form of the building is dictated largely by the energy savings which can be gained with large, simply shaped spaces for reducing the quantity of exterior surface and reduce the possibility of heat loss or heat gain. Also, about 75% of the tower's exterior walls are so well insulated that they are virtually like refrigerator walls. The surfacing of the highly insulated tower is reflective opaque glass, which, in this warm, sunny climate, rejects larger amounts of solar radiation than concrete, stone, aluminum, or various other metal panels. Its light weight eliminates the need for a costly structural foundation, and it is economical in terms of initial cost and maintenance.

Figure 5-11. *Georgia Power Company Building.* (© 1981 by E. Alan McGee, Atlanta.)

Round-the-clock operations are located together so that the entire building need not be conditioned 24 hr a day. In addition, the
mechanical cores and other operations not requiring windows are
located along the east and west walls, providing further buffers to
the sun.

Heery and Heery found that it was "extremely helpful" to establish a maximum energy budget for the building in the predesign phase.
The budget encouraged design to become an integrated response to the
manner in which the whole building worked in relation to its designed use.

Commercial Case Study 2. One of the better known nonresidential
passive projects in the nation is a publications office and warehouse for a Pecos, New Mexico, Benedictine monastary (see Fig. 5-12).
Designed by architect Mike Hansen in conjunction with Steve Baer's
Albuquerque-based Zomeworks firm, the building's 7000-ft^2 single-
level space is enclosed on three sides by sealed (except for summer
cooling vents) 8-in. block construction. Its energy performance
hinges on the 140-ft high upper-level clerestory, which admits sunlight to heat and illuminate the warehouse space at the back of the
building. A middle row of glazing warms the front office space in
the same fashion, also allowing sunlight to strike the heat-retentive
central masonry wall between the spaces.

The lower band of glazing fronts a Trombe wall of 138 fifty-
five gallon water-filled barrels, painted black and stacked within
an insulated interior cabinet that also provides counterspace in the
offices. The thermal storage of the Trombe wall is tapped through
standard counterside registers in each office. Cold air returns,
situated in the flor 2 to 4 ft from the heat registers, facilitate
heat flow through the space.

Commercial Case Study 3. The Copeland, Sinholm, Hagman, Yaw architectural firm designed Aspen's predominantly passive airport terminal in 1975, with the assistance of the Zomeworks firm (see Fig.
5-13). The single-story, 16,800-ft^2 terminal contains offices and
waiting rooms within 8-in. solid concrete block walls which are set

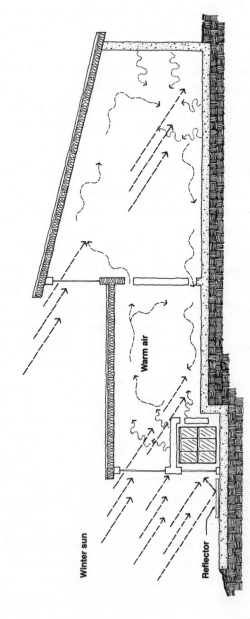

Figure 5-12. Benedictine Publications office and warehouse. (From Ref. 5; courtesy of Steve Baer, Albuquerque, N. M.)

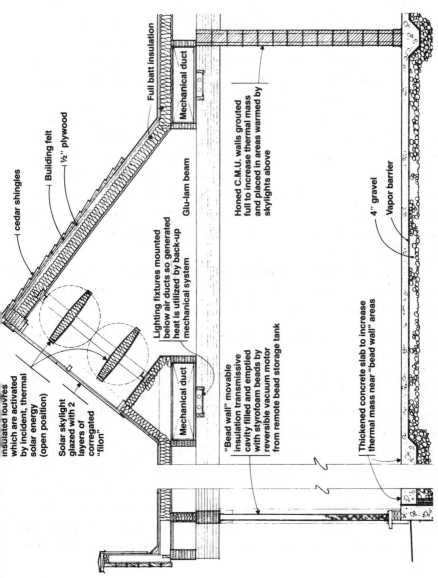

cedar shingles

Building felt

½" plywood

Full batt insulation

Mechanical duct

insulated louvres
which are activated
by incident, thermal
solar energy
(open position)

Solar skylight
glazed with 2
layers of
corrugated
"filon"

Lighting fixtures mounted
below air ducts so generated
heat is utilized by back-up
mechanical system

Glu-lam beam

Honed C.M.U. walls grouted
full to increase thermal mass
and placed in areas warmed by
skylights above

Mechanical duct

"Bead wall" movable
insulation transmissive
cavity filled and emptied
with styrofoam beads by
reversible vacuum motor
from remote bead storage tank

Thickened concrete slab to increase
thermal mass near "bead wall" areas

4" gravel

Vapor barrier

Figure 5-13. Aspen Airport. (From Ref. 1; courtesy of Copland, Hagman, Yaw Ltd., Aspen, Colo., and Dave Harrison, Zomeworks, Albuquerque, N. Mex.)

on a 5-in. concrete slab. The east walls are bermed, and all walls
and ceilings are insulated to an R-20 value.

The staggered southern facades of the building are Beadwall--
a Zomeworks' concept developed by Dave Harrison. It consists of two
translucent fiberglass sheets spaced 2.7 in. apart and filled with
styrofoam beads that can be drawn out to admit solar radiation, or
blown back in at night to provide full-wall insulation.

Solar heat also enters the building through multiple clere-
stories that are oriented to the south and enclose operable insulating louvers called Skylids, another Zomeworks' concept. Actuated
automatically by a solar-powered mechanism, the aluminum-sheathed
louvers open to let sunlight strike the mass of the interior floor
and walls, where heat is stored and radiated back into the interior
space as temperatures drop. The louvers close at night to interpose
5 in. of insulation.

In the cold but sunny Aspen winter, the terminal's hybrid energy
system meets 35 to 40% of its heating requirements; a gas-fired,
forced-air system supplies the difference.

Commercial Case Study 4. The designer has a critical role to play
in the rehabilitation, constructive reuse, and adaptation to a changing environment of our existing commercial building stock, for our
present stock of buildings, especially in our older cities, forms
not only a visible part of our heritage, but also constitutes a man-
made resource worth many times its original cost. What, after all,
is the energy cost of demolishing existing materials and spaces and
replacing them with products yet to be produced? The recent flurry
of restoration projects makes clear that designers are already assuming a leadership role in recycling this resource, not only so
that these structures may be a joy to the human spirit, but so that
they may be persuasive demonstrations of the possibilities of energy-
conscious design.

One of the most successful and sizable examples of renovating
in the past few years has taken place in Boston, Massachusetts, at
Faneuil Hall Marketplace (see Fig. 5-14). With an eye to cutting

Figure 5-14.   Faneuil Hall Marketplace.   (Courtesy of the Rouse Company, Columbia, Md.)

operating costs, architects Ben Thompson and Associates carefully thought out new means of conserving energy by using to advantage features already designed into the original structure. The long, central colonnade was widened by covering the two parallel side streets with glass canopies that can be opened using overhead doors.

In the summer, the market aisles capture natural east-west breezes from the harbor, and the trees in the plaza provide shade. Thus, no air conditioning is needed on the entire first level of the market. In winter, the glass slides above the aisles are closed, and winter solar heat is captured, creating a greenhouse effect.

Midway across the country, the restoration of the Chicago Navy Pier on Grand Avenue at Lake Michigan represents the triumph of imagination over decay. As designed by architect Jerome R. Butler, the restoration also demonstrates how the initial investment required by a project of this magnitude can be justified by the significant decrease in operating expenses which have traditionally been the most costly factor in the whole area of life-cycle costing. Solar collector panels (8000 ft^2) now provide about 35% of the heat to one of the east end buildings. Also, provisions for maximum natural ventilation were made, thanks to the pier's location, 1 mi. out in the lake. In short, the heating, cooling, and ventilating systems save energy and therefore cut costs, while returning a valuable recreation center to use in the Chicago community.

THE SHAPE OF THE FUTURE

As I see it, the question is not whether our built environment will change, but how and when. The limited supply of resources, the economics of man-produced energy, and the government's performance standards, which loom on the horizon, will change and are *already* changing the way we are doing things. However, if we do not modify our response to the problem away from our habitual reliance on *more* technology and from a blind faith in sealed, air-tight structures as the hope for the future, then we can surely expect hardship, damage, and dislocation. Instead, architect and engineer together must face the issue as a team, each contributing his knowledge and professional judgment to a building strategy orchestrated by design. Working together, they can restructure the framework of the energy question so that it appears not as a threat but as a challenge, an opportunity to develop nothing less than a new aesthetic, an aesthetic with a distinctly human, rather than mechanical, bias.

And what will this new aesthetic look like? Different (13). Buildings will be narrower. Facades will be directed to the most desirable orientation. Facades will be more complex and differences in scale and fenestration will vary to reflect different activities.

Walls will once again be capable of admitting or excluding outside air. Such things as operable sashes, sunshades, sunscreens, and horizontal and vertical louvers will become commonplace.

The throwaway emphasis in our culture will give way to adaptive reuse. A widely expanded family of components that perform various operations will be available for insertion into facades. Quality of spaces within buildings will be infinitely more varied and individually personalized, because space will be designed according to use by people.

Buildings will have less reverence for and adherence to symmetry and geometric formalism. However, this will not be a function of the designer's caprice but a more sensitive response to site and climate, a response that may even be mandated by regulations.

The city will still be with us, containing closely grouped dwellings, zero property-line townhouses, and the return of the smaller, expandable-size houses of the 1940s and 1950s. There will, however, be more greenways to provide recreational relief and to offset the heat retention rate of buildings and streets.

Power and heat will be provided through an increasingly varied combination of sources, which will include active and passive solar power, wind-driven generators and pumps, and hydropower electric generators. Energy-producing methods will be kept as small as possible by new attitudes toward the design of buildings. Fossil fuels will be used more as sources for backup rather than primary systems.

What I see is not an aesthetic of *more* of this or *less* of that. It will be an aesthetic of wise choices that resemble, in their economy, the kind of elegance mathematicians find in the solution of a problem, solutions that provide the most direct route and the few-

est number of steps. For architecture, this is the route paved by
design, energy-conscious design.

Going beyond the immediate issue of energy, what the public receives from such an approach is a new regionalism, a sense of place.
But isn't that what our older cities have been saying to us all
along? The unique charm of Savannah, Charleston, New Orleans' Vieux
Carré, Boston's Beacon Hill--aren't these the result of design being
informed by and in turn informing the special nature of a site?
Isn't that why the citizens of these communities have a special
sense of being "at home"--because the buildings are in a sense "living," since they respond to sun, wind, and time?

CONCLUSION

The lessons of the past teach us that design has an important role
to play in what journalists style as today's "energy crisis."
Present research and already functioning energy-conscious structures
tell us that the designer, working together with other professionals,
has a unique contribution to make in preparing for a future of
choices and for an improvement in the quality of life. In this
sense, we stand on the threshold not of crisis but of opportunity
and challenge, a challenge to develop a new aesthetic that, like a
living organism, derives its strength and beauty from the energy of
those who use the built environment--people.

Already the winds of change are blowing. We can turn our backs
to them, but at our peril. Or we can unseal the windows and let the
fresh air come in. Significantly, this has already been done in
Washington, where the transoms in General Meigs' Pension Building
have been reopened and the space openings in the brickwork unstopped.
Present plans call for this structure to house a national museum of
architecture. If these plans are realized, General Meigs' building
will be more than a museum. By reminding us through being an actual
working model that buildings can have comfort designed into them,
more than transoms will have been opened. The fresh air blowing
through will be nothing less than the currents of challenge

and opportunity, currents which, if navigated wisely, will carry us as a profession and as a people into a better designed, energy-conscious tomorrow.

NOTES AND REFERENCES

1. Much of the material in the following discussion of passive technology appeared originally in Res. Design, vol. 2, no. 1, published by AIA Research Corporation, Washington, D.C., January 1979.

2. Hundreds of passive design solutions have been tried by architects. The AIA Research Corporation's (AIA/RC) *Survey of Passive Solar Buildings,* compiled as part of the Passive Solar Systems Study for HUD, documents 100 of these passive solar projects in 28 states.

3. Ralph Knowles, *Energy and Form,* MIT Press, Cambridge, Mass., 1974.

4. Adapted from Res. Design, vol. 2, no. 1, January 1979, p. 6.

5. In the fall of 1976, AIA/RC sponsored and administered a solar dwelling design competition, funded by Exxon Corporation for architectural students in the United States and Canada. For a discussion of this competition, see Res. Design, vol. 1, no. 1, January 1978, p. 12ff.

6. A more detailed analysis of Public Law 94-385, as well as the research project into building energy performance undertaken by AIA/RC for HUD and the Department of Energy, may be found in Res. Design, vol. 1, no. 4, October 1978, p. 9ff.

7. Quoted from the original Request for Compensation Proposal inviting those designers who had participated in Phase 1 of this project to participate in Phase 2.

8. These figures are derived from the Executive Summary (p. 18) of the Phase 2 research project. This information may be obtained from HUD and the Department of Energy.

9. Unless noted otherwise, the case studies that follow are taken from *Passive Solar Heating,* a publication of the Department of Energy, as well as from Res. Design, vol. 2, no. 1, January 1979, p. 7ff.

10. "Trombe wall": Named for Dr. Felix Trombe, one of its developers. As part of a passive heating system, Trombe walls are typically from 12 to 16 in. thick, are on the south side of a structure, and are made of concrete, stone, or masonry. To facilitate their function as heat-storage walls, the exterior surface is usually painted black. Trombe walls may or may not have glazed window openings, but they usually have vents at regular inter-

vals both along the floor and the ceilings of each room. A Trombe wall without vents can be used to store heat for night-time use and stabilize indoor temperatures. But they do not provide much daytime heat.

A Trombe wall with vents performs two important heating functions: (1) it initiates a daytime convection heating loop; (2) at night it allows the radiation of heat stored in the wall. A vented Trombe wall also has summer cooling functions. During the day, the upward convection between glass and concrete wall draws heat out of the house and brings cool air in. At night all vents are opened to encourage the convection loop to reverse, thus cooling the interior.

Adapted from *Homeowner's Guide to Solar Heating,* Sunset Books, Lane Pub. Co., Menlo Park, Calif., 1978.

11. "R-value": The thermal resistance of a material or structure. The higher the R-value, the greater the resistance.

12. George T. Heery, Energy and architecture, Design Quart., published by Heery & Heery, Architects & Engineers, Inc., Atlanta, Ga., Summer 1978.

13. The substance of the observations that follow was suggested by a speech delivered by Herbert Epstein, FAIA, in October 1978 to the Southern Conference for Architects and Engineers.

6

Building Computer Analysis Methodologies

Maurice Gamze

Gamze-Korobkin-Caloger, Inc.
Chicago, Illinois

In his analysis of alternative building environmental control system concepts, an engineer must consider heating and cooling requirements for two distinct time periods. Extreme conditions for an hour or a day must be analyzed to determine whether the desired internal conditions can be maintained during these peak periods, and the typical annual energy requirements must be analyzed to minimize energy costs and/or consumption. To these ends, several manual techniques, such as degree-day and bin methods, have found widespread use in estimating annual energy consumption. More recently, the design engineer has become increasingly reliant on the use of computerized hour-by-hour simulation models for analysis of monthly and annual energy performance. These computerized methods appear to offer such benefits as accuracy, repeatability, speed, and cost. Both state and federal governments have required their use as a means of assuring that the energy performance of buildings will meet specified annual budgets.

Computerized energy analysis systems generally consist of three distinct models. One model is the *loads* model, which computes the hourly heating and cooling requirements by using data determined by

building characteristics, occupancy patterns, and weather patterns.
A second model is the *systems* model, which analyzes various heating,
ventilation, and air conditioning (HVAC) concepts and systems as
their performances in meeting the building loads are simulated. A
third model is the *economics* model, which computes the annual energy
costs according to the energy demand and consumption data produced
by the systems model and the energy cost and rate data. These models
parallel the operation dynamics of actual buildings. For example,
as heat is transmitted through the building skin or introduced internally, the need to either introduce or remove heat in order to
maintain the desired internal temperature is accomplished by the
building HVAC system, which must respond to the changing building
loads. Energy costs are then computed, based on annual, monthly,
or even hourly energy requirements.

How accurate are these computerized techniques as predictors of
a building's energy performance? A traditional measure of verification is the program's ability to accurately predict both typical
monthly and annual energy consumption and peak-day heating and cooling loads. These techniques are very useful in the comparison of
simulated performance to actual performance and in the use of the
technique for sensitivity studies.

With the propagation of computerized techniques, both in the
private and public sectors, various attempts have been made to compare program outputs. Again, the primary measures for comparison
are typical seasonal consumption at specified peak heating and cooling conditions.

The use of computer programs to demonstrate compliance with
state and/or federal building energy design goals has obviated a
requirement for comprehensive, scientific techniques for comparative
analyses of automated energy calculation procedures. Several comparative studies of building energy system simulations have been
performed. In general, differences in output can be attributed to
the following:

1. Mistakes and inaccuracies in handling program output, caused by misinterpretation of output data, differences in output factors, and errors introduced in converting to a common base

2. Mistakes and differences in program input, caused by differing interpretations of available input data, differing assumptions for incomplete or missing data, and errors in program input

3. Internal program differences, including variations in analytic techniques and basic assumptions, and programming and data management errors

The first two factors are primarily functions of the skill of the program user, while the third factor is a function of the program developer. Little information is generally available on a program's internal structure and, in the past, the quantification of the significance of these differences has been most elusive. The quantification is the subject of this chapter.

LOAD MODELING

Many factors influence the energy performance of a building. Before engineers can utilize the various computerized techniques to demonstrate compliance to energy standards, they must first understand how each computer program models these factors and analyzes their respective impact on energy and cost performance.

The first step required in the modeling of a building's energy performance is the computation of the building's heating and cooling loads. The heating and cooling requirements of a building are functions of many factors:

1. The difference between the temperature of the outside air and the temperature of the conditioned space

2. The solar radiation and heat load on the building

3. Internal, sensible, and latent heat gains from people

4. Internal, sensible, and radiative heat gains from both equipment (e.g., lighting) and building internal surfaces.

The heat flows through a building's outside walls are shown in Fig. 6-1. In this illustration

x = distance from inside wall surface measured toward outside wall

$T_{o,a}$ = outside air temperature

$T_{i,a}$ = inside air temperature

$T_{i,s}$ = wall, inside surface temperature at $x = 0$

$T_{o,s}$ = wall, outside surface temperature at $x = L$

Q_R = solar radiation heat gain

Q_L = lighting heat gain

Q_P = sensible and latent heat gain from people

Q_S = internal heat gain from other surfaces

The many load computer programs that are available on the market model each of these heat flows by using different assumptions regarding their impact and interaction and by using different computational algorithms. Table 6-1 is a tabulation of the models utilized in the 18 different load computer programs examined. A review of the individual models indicates significant differences both among the models and among their use within individual load programs. It is also apparent that these model differences will affect load computations differently by varying as functions of the building's characteristics. From an operational perspective, one is faced with

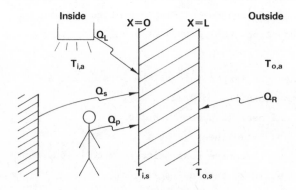

Figure 6-1. Building heat flow.

Table 6-1a. Characteristics of Load Programs

	Julotta (Sweden)	Atkool (Britain)	ZKWF31 (Germany)	LPB-1 (Belgium)	DOE-2 (USA)	ESP (Abacus) (Britain)
Heat Transmission Through Opaque Wall	Numerical Solution of Fourier Eq.	Response Factor Method	Transfer Matrices to Fourth Harmonic Order	Response Factor Method	Response Factor Method	Numerical Solution of Fourier Equation
Solar Effects on Opaque Walls	Sol-Air Temperature	Sol-Air Temperature	Sol-Air Temperature	RFM: Rectangular Pulse Excitation	Sol-Air Temperature	Heat Balance at Exterior Surface
Cooling Load from Heat Transmission Through Opaque Wall	Numerical Solution of Heat Balance Equation	Room Transfer Function	Numerical Solution of Heat Balance Equation	Steady State (No Time Lag)	Transfer Function Method	Numerical Solution of Heat Balance Equation
Split of Indoor Film Coefficient	Convective Separate from Radiative	No Split	Convective and Computed Effective Radiative	Convective Only	No Split	Convective (Horizontal and Vertical) Separate from Radiative
Modeling of Long Wave Radiation Exchange	Detailed Modeling	None	None	None	None	Detailed Modeling
Instantaneous Incident Solar Gains Through Glass	Electromagnetic Concepts	Absorption Transmission Polynomial Coeff. for Solar Heat Gains	Transmitted Radiation. Computed with Empirical Formula of Angle θ	None	A-T Poly. Coeff. for Solar Heat Gain or Shading Coeff.	Detailed Analysis with Interior Geom. Taken into Consideration
Transmitted Solar Radiation Distribution in Interior	All Diffuse, Total Exchange Factors and Reflective Coeff. Used	No Modeling	No Modeling	No Modeling	No Modeling	Determined by Projection of Window onto Internal Surfaces
Cooling Load from Solar Heat Gains	Numerical Solution of Heat Balance Eq.	Room Transfer Functions	ASHRAE Simplified Method	Thermal Response Factor Method	Transfer Function Method	Numerical Solution Taken into Account Interior Insolation Pattern
Modeling of HVAC Systems and Equipment	None	None	None	None	Models Most Air Systems	None

Table 6-1b

	HTB (UWIST) (Britain)	Airnet (Scotland)	ESA (USA)	GEPVP (Britain)	Scout (USA)
Heat Transmission Through Opaque Wall	Numerical Solution of Fourier Equation	Numerical Solution of Fourier Equation	Steady State Method	Numerical Solution of Fourier Eq.	Response Factor Method
Solar Effects on Opaque Walls	Heat Balance at Exterior Surface	Heat Balance at Exterior Surface	Total Equivalent Temperature Differential	Sol-Air Temperature	Sol-Air Temperature
Cooling Load from Heat Transmission Through Opaque Wall	Numerical Solution of Heat Balance Equation	Numerical Solution of Heat Balance Equation	Steady-State (No Time Lag)	Thermal Response Factor Method	Transfer Function Method
Split of Indoor Film Coefficient	Convective Separate from Radiative	Convective Separate from Radiative	No Split	No Split	No Split
Modeling of Long Wave Radiation Exchange	None. Mean Radiant Temperature is Used	Detailed Modeling	None	None	None
Instantaneous Incident Solar Gains Through Glass	Using Refractive Index, Absorption Coeff., Glass Thickness	Function of Angle o, Absorption Coeff., and Glass Thickness	Shading Coeff.	Shading Coeff.	Shading Coeff.
Transmitted Solar Radiation Distribution in Interior	All Shortwave Solar Radiation is Absorbed by the Floor	All Direct Absorbed by Floor. Diffuse Uniformly Distributed	No Modeling	All Diffuse. Angle Factors and Distribution Coeff. Used	No Modeling
Cooling Load from Solar Heat Gains	Numerical Solution of Heat Balance Equation	Numerical Solution of Heat Balance Equations	ASHRAE Simplified Method	Thermal Response Factor Method	Transfer Function Method
Modeling of HVAC Systems and Equipment	None	None	Models Most Air Systems	None	Models Some Air Systems

Table 6-1c

	RCPL (Canada)	Ventac (Sweden)	Therm (Britain)	WTE-01 (Holland)	BA-4 (Denmark)	Faber (Britain)
Heat Transmission Through Opaque Wall	Transfer Function Method	Numerical Solution of Fourier Equation	Response Factor Method	Numerical Solution of Fourier Equation	Steady State Method	Response Factor Method
Solar Effects on Opaque Walls	Heat Balance at Exterior Surface	Heat Balance Exterior Surface	Heat Balance at Exterior Surface	Heat Balance at Exterior Surface	Empirical Method Function of U Value of Wall	Sol-Air Temperature
Cooling Load from Heat Transmission Through Opaque Wall	Room Transfer Function	Numerical Solution of Heat Balance Equation	Numerical Solution of Heat Balance Equation	Numerical Solution of Heat Balance Equation	Numerical Solution of Heat Balance Equation	Transfer Function Method
Split of Indoor Film Coefficient	No Split	Convective Separate from Radiative	Convective Separate from Radiative	Convective Separate from Radiative	Convective Separate from Radiative	No Split
Modeling of Long Wave Radiation Exchange	None	Detailed Modeling	Detailed Modeling	Detailed Modeling	None	None
Instantaneous Incident Solar Gains Through Glass	A—T Polynomial Coeff. for Solar Heat Gains	A—T Polynomial Coeff. for Transmitted and Absorbed Radiation	A—T Polynomial Coeff. for Transmitted and Absorbed Radiation	Shading Coeff. Based on ¼ Inch Solar Grey Glass	Shading Coeff. Based on Clear Double Pane Glass	Using Transmittance Absorptance, Reflectance at Normal Incidence
Transmitted Solar Radiation Distribution in Interior	No Modeling	Determined by Total Exchange Factors	90% on Floor 10% Evenly Distributed Diffuse Evenly Distributed	Uniformly Spread Over All Interior Surfaces	Evenly Distributed Over All Surfaces	No Modeling
Cooling Load from Solar Heat Gains	Transfer Function Method	Numerical Solution of Heat Balance Equation	Numerical Solution of Heat Balance Equation	Numerical Solution of Heat Balance Equations	Numerical Solution of Heat Balance Equation	Transfer Function Method
Modeling of HVAC Systems and Equipment	None	Models Some Air Systems	None	None	None	None

the fact that even if input and output differences are eliminated,
because of differences in their internal structure, load programs
will produce different estimates of heating and cooling loads for
the same building characteristics. Which program is "best" for
modeling any specific building? To generate the information required to answer this question, 18 load programs were exercised,
with input obtained from a carefully developed, detailed set of
building specifications. These specifications, developed for a set
of hypothetical buildings, were used to identify the varying energy
requirement predictions caused by internal differences in the models.

PROGRAM ANALYSES

To identify internal program differences, input variations were minimized by specifying the physical and thermal parameters of a hypothetical set of buildings (see Table 6-2 and Fig. 6-2). Three varia-

Table 6-2. Hypothetical Building Characteristics

Shape/Orientation	Rectangular with north facade facing 30° west of true north.
Height	12 occupied typical floors resting on raised columns.
Gross Areas (Outside Dimensions)	Building total of 225504 sq. ft. (12 typical floors at 18,792 sq. ft. each).
Structure	Steel frame with 4 inch concrete floor slabs.
Typical Story	261 ft. long and 72 ft. wide, 13.5 ft. floor-to-floor.
Walls	Glass face curtain wall with 1 inch insulation on all walls.
Typical Floor	Dark color carpet on 4 inch concrete slab.
Second (Bottom) Floor	Carpet on 4 inch concrete slab, dead air space and 4 inch insulation on metal lath and plaster ceiling.
Roof	Insulated built-up roof on 4 inch concrete slab.
Window Glass	53% of typical floor facade, ¼ inch solargrey single glaze mounted flush to outside wall face.

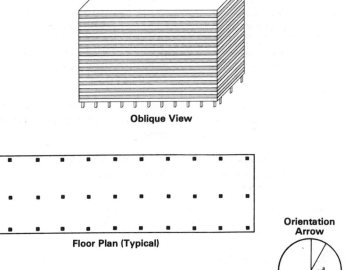

Figure 6-2. Hypothetical IEA building.

tions of the building, as described in Table 6-3, were simulated. Gamze-Korobkin-Caloger prepared the building specifications and experiment design, but the individual programs (Table 6-4) were tested by the people who had developed each program. A series of parametric studies was also performed. Thus, because the program input was rigorously controlled, the differences in output could be attributed to program internal differences.

Each of the 18 programs was tested by using the hypothetical building specifications and by using artificial weather conditions. The following phenomena were explicitly examined:

Building thermal storage effects

Heat exchange between interior surfaces

Convective and radiative components of heat gains from lights and
    people

Table 6-3. Building Variations

Option	Description	Significance
A	No Windows No Internal Loads	Assess techniques for modeling heat transmission and solar gains through opaque surfaces.
B	Windows No Internal Loads	Assess techniques for modeling heat transmission and solar gains through glass.
C	Windows Internal Loads - Lighting and People	Assess techniques for modeling internal heat gains.

Table 6-4. Programs Analyzed

Therm (UK)	Abacus (UK)
Ecube 75 (USA)	Julotta (Sweden)
BA4 (Denmark)	Pilkington (UK)
Meriwether (USA)	Atkool (UK)
DOE - 1 (USA)	WTE-01 (Holland)
Reid (Canada)	Airnet (UK)
Scout (USA)	Ventac (Sweden)
Ecrc (UK)	HTB (UK)
Faber (UK)	
HVAC5 (UK)	

In assessing the program output, the total annual energy consumption, annual and peak heating loads, and annual and peak cooling loads of each computer simulation were reviewed. It should be noted that these loads do not include the effects of ventilation air, infiltration/exfiltration, and HVAC system performance.

During the course of the study, it was found that it was necessary to distinguish between the various programs in the modeling of steady-state effects. To assess the significance of these effects, an evaluative technique, which is based on the fact that the net monthly thermal requirement (heating load minus cooling load) is independent of a building's non-steady-state or storage effects, was developed. This technique is based on the fact that since, over a

1-month period, the net change in the heat stored in a wall is very small, compared to the heat transferred through the wall, the net heat transfer will be proportional to the differences between the interior temperature and the sol-air or effective outside temperature. A regression technique was then used on simulated loads in order to determine the impact in the modeling of building construction transient effects.

Modeling of Building Thermal Storage Effects

The basic computation techniques consist of the following:

1. Explicit modeling of the building and the detailed *numerical solution* to the partial differential equations which describe heat transfer in the building

2. A *response factor* approach, which also requires the explicit modeling of the specific building's interior and in which loads are computed as functions of temperature differences, solar radiation, and internal gains

3. *Predetermined transfer* functions for similar buildings, in a response factor algorithm, which reduces the need for explicit interior modeling

4. Various *simplified techniques*

The technique used to model the building thermal storage effects is the major cause of differences in both the annual heating and cooling requirements and the annual heating and cooling peak demands. Programs which used the American Society of Heating, Refrigerating and Air Conditioning Engineers (ASHRAE) simplified method (1), for example, overestimated the annual heating and cooling requirements by 56%, the annual heating peak demand by 15%, and the annual cooling peak demand by 27%, with respect to programs which used the response factor method. The magnitudes, causes, and complications of these differences are discussed in the sections that follow.

Magnitudes of Load Differences. For building variation C, programs DOE-1 and SCOUT yielded similar results, but these results differed substantially from programs ESA and ECUBE 75. Annual loads computed by these four programs are summarized in Table 6-5. Figure 6-3

Table 6-5. Comparison of Annual Heating and
Cooling Requirements (Gigajoules)

	Total Heating Req't	Total Cooling Req't	Annual Heating Peak	Annual Cooling Peak
Ecube 75	7112 (159)	11,523 (155)	3638 (123)	6547 (125)
ESA	7294 (163)	10,738 (144)	3611 (120)	6932 (132)
DOE-1	4507 (101)	7673 (103)	3061 (102)	5162 (98)
Scout	4464 (100)	7454 (100)	2999 (100)	5252 (100)

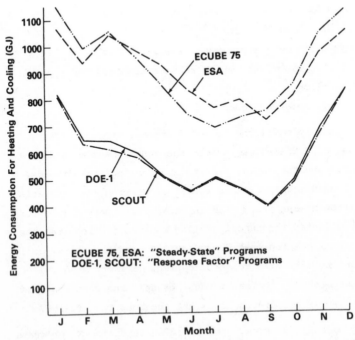

Figure 6-3. Total monthly energy consumption for heating and cooling in building (with internal loads).

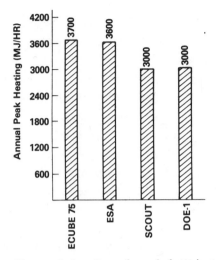

Figure 6-4. Annual peak heating demand in building (with internal loads).

shows the monthly energy consumption obtained by adding the heating requirements to one-third of the cooling requirements. Histograms in Fig. 6-4 and 6-5 show the annual heating and cooling peaks, respectively.

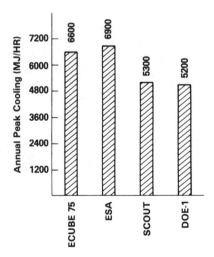

Figure 6-5. Annual peak cooling demand in building (with internal loads).

Causes of Load Differences. The differences in the annual energy
consumption, the annual peak heating demand, and the annual peak
cooling demand computed by the four programs result entirely from
differences in modeling of the building's thermal inertia. The long-
term effects indicated by the values of the *net* thermal loads (heating minus cooling) are essentially similar, as shown in Fig. 6-6.
Moreover, Fig. 6-7 indicates that while the value of heating minus
cooling are essentially similar for the cyclic summer day (clear
sky), the profiles are distinctly different. Programs ECUBE 75 and
ESA predict a heating load during the night and a cooling peak demand of about 7500 MJ/hr during the early afternoon. However, programs DOE-1 and SCOUT predict a small cooling load during the night

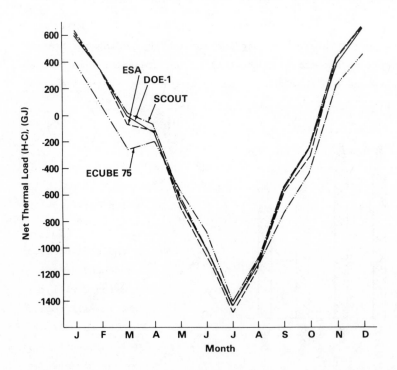

*Figure 6-6. Net thermal load (heating-cooling) in building (with
internal loads).*

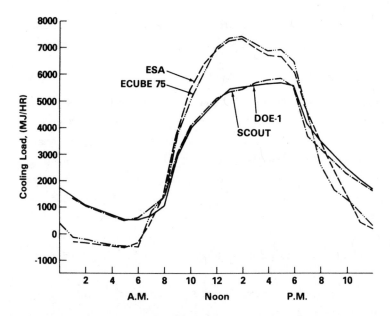

Figure 6-7. Hourly load profiles of building for cyclic summer day (clear sky) (heating-cooling).

and a cooling peak demand of less than 6000 MJ/hr occurring during the late afternoon. Although all of the four programs predicted similar net energy flows across the building shell over the 24-hr period, they predicted differing energy consumption and peak demands. These differences in demands reflected the differences in modeling of the thermal storage effects of the building, which determines the hour-by-hour variations of the heating or cooling load.

Implications of Load Differences. Sizing heating and cooling equipment should depend on the peak heating and cooling demand. For example, based on the DOE-1 program, building variation C will require a 410-ton chiller. Based on the ESA program, it will require a 545-ton chiller. This implies that the initial cost as well as the operating cost predicted by the ESA program are substantially larger than those predicted by the DOE-1 program. The differences in predicted performance are important for establishing building

energy performance standards and for studies involving the relationship between building architecture and building energy performance.

Modeling of Interior Surface Heat Exchange Effects

The four basic storage modeling options described previously can be
used for analyzing the heat exchange effects among interior surfaces.
Except for those that used the simplified ASHRAE method, programs
that adhered very closely to the basic building specifications predicted similar values of annual energy consumption in building variation C. As long as the critical parameters specified were adhered
to, modeling the heat exchange between the interior surfaces had
little effect on the annual energy consumption. However, there was
a noticeable impact on the predicted annual peak cooling load in
building variation C.

Programs that used transfer function or response factor techniques without modeling the interior radiation distribution or interior surface heat exchange overpredicted the three major components
of the cooling load. These programs overestimated the solar heat
gains through glass by about 20%, the heat transmission through
glass as a result of temperature difference by about 15%, and the
internal heat gains from lights by about 7%. These programs also
overestimated the component of the heating load, caused by heat
transmission loss through glass, by about 15%.

The net effect was a simultaneous decrease in the heating load
by about 5% and increase in the cooling load by about 3%. The overestimate in the heat transmission loss through glass was nearly balanced by the overestimate in the solar and internal heat gains.
Modeling the interior surface resulted in a small change in the total
energy consumption for heating and cooling of building variation C.

Magnitudes of Load Differences. Figure 6-8 shows the total monthly
energy consumption for heating and cooling of building variation C,
computed by programs JULOTTA and DOE-1. The computed annual heating and cooling requirements were:

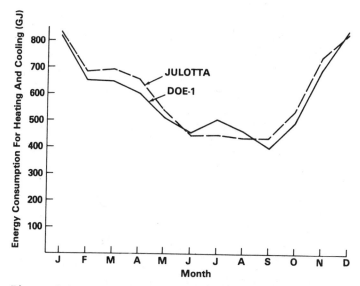

Figure 6-8. Total monthly energy consumption for heating and cooling in building. Annual energy consumption (GJ): JULOTTA, 7256; DOE-1, 7065. JULOTTA: Modeled interior radiation exchange. DOE-1: Did not model interior radiation exchange.

	Heating Requirement (GJ)	Cooling Requirement (GJ)
JULOTTA	4746	7530
DOE-1	4507	7673

Figures 6-9 and 6-10 show histograms of the annual peak heating demands and annual peak cooling demands, respectively. The two program predictions are noticeably different for peak demands for cooling.

Causes of Load Differences. Programs that use transfer function or response factor methods without modeling the interior follow ASHRAE procedures for calculating heating and cooling loads. These procedures, which are described in Chapters 25 and 26 of the *1977 ASHRAE Handbook of Fundamentals,* do not adequately and comprehensively account for all energy eventually reflected to the outside through the

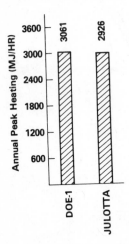

Figure 6-9. Annual peak heating demand in building (with internal loads).

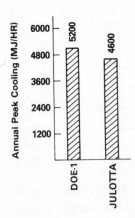

Figure 6-10. Annual peak cooling demand in building (with internal loads).

building shell. Such a comprehensive accounting requires detailed modeling of the indoor longwave radiation exchange and of the short-wave reflection to the outside. Moreover, detailed modeling of the interior gives a more accurate estimate of the heat transmission losses through the glass surfaces as a result of temperature differences.

The components of the thermal loads which caused differences in the results of programs DOE-1 and JULOTTA are described in the following sections.

Heat Transmission Through Glass as a Result of Temperature Differences. Table 6-6 presents the computed heat transfer through glass on a cyclic winter day which has a constant ambient dry bulb temperature of $20^{\circ}F$ and no solar insulation. Heat transmission is defined as:

$$\text{Heat transmission} = H_B - 0.55 \ H_A \qquad (1)$$

where

    H_B = total heating load for building variation B

    H_A = total heating load for building variation A

The difference between H_B and $0.55 \ H_A$ represents the heating load caused by transmission heat loss through glass surfaces.

The winter cyclic (repeating) day results are significant because they do not include the effects of transients. The heating load on the winter steady-state day is a direct measure of the amount of energy lost through the glass because of temperature differences.

Table 6-6. Amount of Heat Transmitted Through Glass Due to Temperature Differences on Cyclic (Repeating) Days

	Heat Transmission (MJ)	%
DOE-1	75,384	100
Julotta	65,400	87

JULOTTA predicted lower heat transmission through glass than did
DOE-1. The lower values assigned by JULOTTA for the term H_B - 0.55
H_A shows the effect of modeling the interior on the predicted heat
transmission losses through glass surfaces. The differences in predictions between the two programs is caused by splitting the indoor
film coefficient into radiative and convective portions.

Convective heat transmission through glass is a function of the
inside air-outside air temperature difference. The radiative transfer of heat is a function of the hot surface-cold surface temperature difference. The temperature of the inside surface of the walls
is slightly less than $70^{\circ}F$, while the temperature of the glass is
much lower than $70^{\circ}F$ (on the winter steady-state or repeating day).

Net Space Solar Energy Gains Through Glass. The results of the analysis of the cyclic summer day (clear sky, June 1 solar values, ambient temperature between 63 and $83^{\circ}F$) are shown in Table 6-7, which
also shows the net space solar energy gain [Eq. (2)].

$$\text{Heat gain} = (C_B - 0.55C_A) - (A_B - 0.55A_A)\left(\frac{DH_C - DH_H}{24}\right) \qquad (2)$$

where

C_A = cooling load for variation A
C_B = cooling load for variation B
DH_C = cooling degree-hours
DH_H = heating degree-hours
and $(A_B - 0.55A_A)$ is proportional to the total glass transmittance, $i^A{}_{ig}U_{ig}$

Table 6-7. *Amount of Net Space
Solar Energy Gain Through Glass
on Cyclic Summer Day*

	Heat Gain (MJ)	%
DOE-1	37,340	100
Julotta	30,709	82

The net solar heat gains predicted by JULOTTA are 82% of those predicted by the DOE-1 program for the cyclic summer day. The reason for this difference is the various methodologies of the programs. Program JULOTTA is based on the detailed analysis of the interior longwave radiation exchange and shortwave radiation distribution; the DOE-1 program is not.

The heat gains through glass assumed four forms:

1. Heat gain by longwave radiation (direct and diffused solar radiation absorbed by glass, then released into space by longwave radiation)

2. Heat gain by convection (absorbed energy released into room space by convection)

3. Direct shortwave-transmitted radiation

4. Diffuse shortwave-transmitted radiation

Program JULOTTA distinguishes between these four forms of heat gain through glass and uses shape factors and distribution factors to perform a detailed heat balance in the rooms. On the other hand, program DOE-1 does not distinguish between the above forms of heat gains and does not perform a detailed heat balance which takes into account the radiative heat exchange within the room. The outcome of the difference in methodology between JULOTTA and DOE-1 was that the latter overpredicted the net space solar heat gain through glass for the cyclic summer day by about 20%. This finding was verified through manual computations, which demonstrated how the detailed analysis of the longwave radiative heat exchange used to compute the indoor surface temperatures and the cooling loads can result in a net solar heat gain through glass which is only 76% of the "ASHRAE value."

Internal Heat Gains from Lights. The total internal heat gains C_{INT} for the cyclic summer day were calculated by using the following equation:

$$C_{INT} = (C - H)_C - (C - H)_B \tag{3}$$

where

$(C - H)_A$ = net thermal load for building variation C

$(C - H)_B$ = net thermal load for building variation B

Computer values of C_{INT} are 29,411 and 27,699 for DOE-1 and JULOTTA, respectively.

These results indicate that detailed simulation of the interior radiation distribution indeed accounts for the energy reflected to the outside through the building shell. In this case, the portion reflected to the outside is 6% of the total internal heat gains.

Implications of Load Differences. It is important to bear in mind that these findings apply only to the basic building. The differences may become smaller for buildings with smaller glass surfaces, for which the weighted-average shape factor applied to radiant heat transfer to glass becomes closer to unity. These differences become important in the analysis of buildings which utilize passive solar systems. Buildings with large glass surfaces and greenhouses cannot be accurately simulated without accounting for interior reradiation and reflection. The ASHRAE procedures described in the *1977 ASHRAE Handbook of Fundamentals* cannot yield precise loads for buildings with large glass surfaces; as a result, these procedures will theoretically overpredict the peak cooling load.

Modeling of Indoor Parameters

Indoor heat gains are primarily the result of the heat generated by lighting and people. Variations in the manner in which the radiative and convective fractions of these heat gains are modeled can result in significant differences in computed loads. Another parameter that created noticeable differences in results was the split between the convective and radiative portions of the indoor surface film coefficient.

Magnitude of Load Differences. To illustrate the magnitude of the differences created by the convective-radiative split of the internal heat gains, results of programs JULOTTA and HTB are discussed in detail. These two programs yielded similar results pertinent to the building variation B (no internal heat gains), yet differed substantially in modeling of building variation C (with internal heat gains). The annual loads computed by the two programs are shown in Table 6-8. The solar and heat transmission effects were accounted for in similar manners for variation B, while for variation C the effects reflect the different methods for modeling the internal heat gains. Program HTB predicted higher heating and cooling requirements than those predicted by JULOTTA. The internal heat gains from lights for JULOTTA were specified as 50% radiant and 50% convective; however, program HTB assumed that the internal heat gains were 100% convective. The hourly profiles of the cooling loads caused by the internal heat gains from lights are shown in Fig. 6-11. These profiles were obtained by subtracting the hour-by-hour cooling loads of building variation B from the cooling loads of building variation C. Because of the differing profile shapes, program HTB predicted higher annual heating and cooling requirements than those predicted by program JULOTTA.

Causes of the Load Differences. Programs JULOTTA, BA4, THERM, and WTE-01 computed the cooling and heating loads by monitoring the longwave and shortwave radiation distributions and all indoor surface temperatures. The cooling and heating load was computed by using a convective indoor film coefficient. All four programs assumed a 50% split between the radiative and convective portions of the internal heat gains. However, the radiative-convective splits of the

Table 6-8. Annual Heating and Cooling Loads

Building Variation	Program	Heating Req't (GJ)	Cooling Req't (GJ)	Heating Peak (MJ/HR)	Cooling Peak (MJ/HR)
B	Julotta	8347	1038	3043	2027
	HTB	8150	1072	2881	2003
C	Julotta	4746	7530	2926	4583
	HTB	5573	9234	2881	5053

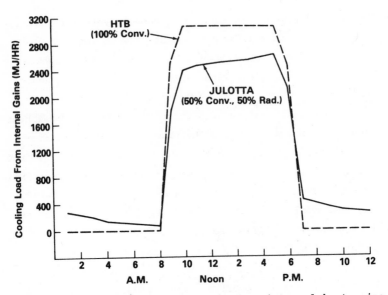

Figure 6-11. *Hourly cooling loads from internal heat gains.*

solar heat gains and of the indoor surface film coefficients varied
to some extent across these four programs.

The output of the four programs is summarized in Table 6-9.

Figures 6-12 through 6-14 show the monthly energy consumptions
and the peak heating and cooling demands. Noticeable differences
occur between the four programs. Since all four programs used the
same convective-radiative split of the internal heat gains, the differences in results may be better explained by examing the loads

Table 6-9. *Annual Loads for Building Variation C*

	Heating Req't (GJ)	Cooling Req't (GJ)	Heating Peak (MJ/HR)	Cooling Peak (MJ/HR)
BA4	4418	7129	2746	4375
Julotta	4746	7530	2926	4583
Therm	5098	6487	3157	4356
WTE-01	5288	7527	3133	5020

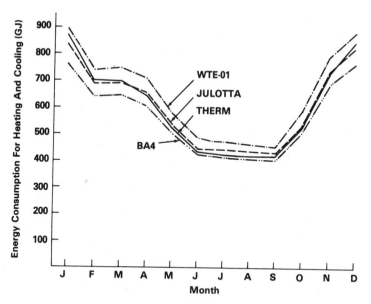

Figure 6-12. Total monthly energy consumption for heating and cooling in building. Annual energy consumption (GJ): BA4, 6794; JULOTTA, 7256; THERM, 7260; WTE-01, 7797.

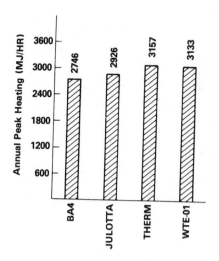

Figure 6-13. Annual peak heating demand in building (with internal loads).

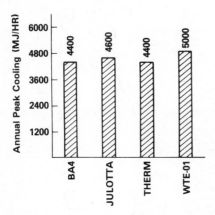

Figure 6-14. Annual Peak cooling demand in building (with internal loads).

for building variation B, as shown in Table 6-10. The main components of the heating and cooling loads in this case are the solar heat gains through glass and the heat transmission through glass. The convective-radiative splits of the four programs are described in Table 6-11.

To analyze these differences, transmission of heat through glass, which is attributed to temperature differences, will be treated separately from the effects of solar radiation incident on the outside surfaces of the glass. The method of analysis will permit this distinction, although the four programs do not keep track of the two effects separately. The effects of temperature difference on the

Table 6-10. Annual Loads for Building Variation B

	Heating Req't (GJ)	Cooling Req't (GJ)	Heating Peak (MJ/HR)	Cooling Peak (MJ/HR)
BA4	8056	826	2850	1801
Julotta	8347	1038	3043	2027
Therm	9247	802	3285	1866
WTE-01	9008	1214	3248	2419

Table 6-11. Program Methodologies

Program	Convective-Radiative Split International Units, W/M², °C	Solar Radiation Distribution
BA4	Transmission through glass Radiative Solar through glass 25% Conv., 75% Rad. Equal Indoor Opaque Surface Temperature	—
Julotta	h_c = 2.99 All surfaces except top floor roof (h_c = 2.07) h_r = 5.40	Uniform Distribution
Therm	h_c = 3.0 for walls, h_r = 5.0 h_c = 1.5 for floors, h_r = 5.0 h_c = 4.3 for ceilings, h_r = 5.0	Trans: 90% on flloors 10% Other Diff: Uniform
WTE-01	h_c = 3.0 h_c = 6.0	10% to Air 90% Uniform

transmission of heat through glass, as estimated by the programs, are computed from the winter repeating-day loads (no solar radiation). The net cooling requirements of the cyclic summer day are used to compute the effects of solar radiation incident on the glass surfaces.

Heat Transmission Through Glass as a Result of Temperature Differences. The results of the analysis of the cyclic winter day (constant ambient dry bulb temperature of $20°F$ and no solar radiation) are shown in Table 6-12, in which heat transmission is defined as in Eq. (1). As a result of the radiative-convective split of the outward heat transfer through the glass surfaces, BA4 predicted 14% lower heat losses through glass than did WTE-01. Program BA4 assumed all transmission through glass to be entirely radiative, while WTE-01 assumed the same to be partly convective (h_c = 3.0 W/m²/°C) and partly radiative (h_r = 6.0 W/m²/°C). Moreover, we observe that programs THERM and JULOTTA yielded essentially similar transmission losses because of the similiarity in the manner in which the indoor

Table 6-12. Amount of Heat
Transmitted Through Glass Due
to Temperature Differences on
Winter Repeating Day

	Heat Transmission (MJ)	%
BA4	60,288	86
Julotta	65,400	93
Therm	64,176	92
WTE-01	69,984	100

film coefficient was split between radiative and convective portions. We also observe that program WTE-01 yielded a larger value for transmission losses because the program analyst used a slightly larger radiative film coefficient. A complete list of the effect of the convective-radiative split on the transmission of heat through glass is shown in Table 6-13. If the value of $(h_c + h_r)$ were kept constant, the higher value of h_r would result in a lower rate of heat transmission through glass.

Net Space Solar Energy Gains Through Glass. The results of the analysis of the cyclic summer day (clear sky, June 1 solar values, ambient temperature between 63 and 83°F) are shown in Table 6-14. The heat gain is as previously defined by Eq. (2).

The largest difference in the net space solar energy gains through glass occurred between programs THERM and WTE-01. Program THERM assumed that 90% of the transmitted solar radiation was indident on the floor, for which h_c was equal to 1.5 W/m^2/°C. On the other hand, program WTE-01 assumed that 10% of the transmitted radiation would be absorbed by air (and seen as an instantaneous cooling load) and 90% would be uniformly distributed over interior surfaces, for which $h_c = 3.0$ W/m^2/°C. Consequently, program WTE-01 predicted higher net space solar energy gains through glass than did program THERM. Moreover, program WTE-01 predicted higher values than

Table 6-13. Relationship Between Heat Transmission Through Glass Due to Temperature Differences and the Convective-Radiative Split of the Indoor Film Coefficient

Program Name	Inside Film Coefficient Convective[e]	Radiative[e]	(Winter Repeating Day Calc.)
	(Btu/Hr/°F/ft²)		
1) Therm	0.53	0.88	15% Lower
2) Ecube 75	1.46[a]		Agreement
3) BA4	0.00	1.46	20% Lower
4) ESA	1.46		Agreement
5) DOE-1	1.46		Agreement
6) RCPL	1.46		Agreement
7) Scout	1.46		Agreement
8) ECRC	M[b]		Agreement
9) Faber	M		Agreement
10) HVAC5	M		Agreement
11) Abacus	M		Agreement
12) Julotta	0.53	0.95	13% Lower
13) Pilkington	1.46		Agreement
14) Atkool	M		Large Deviation[d]
15) WTE-01	0.53	1.06	7% Lower
16) HTB	0.54	0.92	16% Lower
17) Ventac	0.53	0.93	15% Lower
18) Airnet	1.46	0.93	Large Deviation[c]

[a]Where only one number is indicated, it is implied that the program does not split the indoor film coefficient into a radiative and a convective portion.

[b]Information has not been submitted regarding the program methodology.

[c]Deviation was due to misinterpretation of the specification by assuming h_c = 1.46 and h_r = 0.93.

[d]Reasons for the large deviation have not been identified because of Atkool program analysts have not as yet responded to the questionnaires.

[e]$h_r = \varepsilon \, (0.173) \times 4 \, (T/100)^3 \, 1/100 = 0.9 \times 0.173 \times 4 \, (5.3)^3 \, 1/100 = 0.927$ Btu/hr/°F. $h_c = 1.46 = 0.93 = 0.53$ Btu/hr/°F/ft² (from IEA-0 Specifications).

*Table 6-14. Amount of Net
Space Solar Energy Gains
Through Glass on Cyclic
Summer Day*

	Heat Gain (MJ)	%
BA4	26,813	80
Julotta	30,709	92
Therm	24,082	72
WTE-01	33,486	100

did program JULOTTA because of the higher value of h_r and because
10% of the transmitted radiation was directly absorbed by air.

Referring back to Table 6-10 and comparing results of JULOTTA
and THERM, we observe that THERM predicted a higher heating requirement and a lower cooling requirement than did JULOTTA. This is a
consequence of the fact that THERM predicted lower net space solar
energy gains through glass than those of JULOTTA as described previously. A comparison of programs BA4 and WTE-01 shows that as a
result of lower predicting of heat transmission losses through glass
and of net space solar energy gains, program BA4 predicted lower
heating and cooling requirements than did WTE-01.

Implications of Load Differences. The convective-radiative split
of the interior surface film coefficient caused noticeable differences in both heating and cooling requirements and heating and cooling peak demands. In the basic building specifications, the total
value $(h_c + h_r)$ was specified. The radiative portion (h_r) can be
calculated by assuming a constant indoor opaque surface temperature
of $70^{0}F$ and an emissivity of 0.9. The value of h_c can then be derived as the difference between the total specified value and the
computed value of h_r. In reality, h_c depends on the air circulation
within the room and on the dirdction of heat flow. For precise
load simulations, the value of h_c should be estimated accurately;
otherwise, detailed modeling of the interior would not be advantageous.

The impact of air circulation on h_c cannot be overemphasized, particularly when terminal HVAC equipment discharges air close to the indoor surfaces. Induction and fan-coil units placed close to windows will cause an increase of h_c at the inner glass surfaces, and ceiling diffusers will cause an increase in h_c at the ceiling inner surface. Hence, detailed modeling of the interior will be beneficial only if accurate estimates of h_c are utilized.

CONCLUSION

Both peak heating and cooling requirements and annual energy requirements are important outputs of computerized load models. Knowledge of the differences among programs and of the causes of such differences is an important requirement for the effective use of such programs. The major cause of differences in computed energy requirements are attributed to differences in the techniques utilized to model internal heat balances and the building's storage effects. Available techniques range from detailed solutions to the heat flow partial differential equations to various simplified approximations of the heat transfer model.

Simplified techniques for modeling storage and internal radiation effects result in higher load predictions than techniques that utilize more accurate numerical analysis. Thus, for "heavy" buildings for which these effects are significant, the more sophisticated program is appropriate. For "lighter" building construction, less sophisticated techniques which offer other benefits are acceptable.

Transfer function techniques overestimate the heat gain from solar loads, and this in turn increases the calculated cooling load and reduces the heating load. Because of the compensating effect of these differences, the net annual energy consumption is approximately equal to that predicted by more complex techniques. The impact of their differences is on equipment sizing and on the resulting operating requirements; it is most significant in buildings with a large percentage of glass area.

Programs also differ in the manner in which they model the radiative and convective fractions of the internal heat gain. Accurate modeling of these fractions, possible with many detailed numerical techniques, requires detailed information on the building's interior. Unless these data are available, the resulting assumptions that are required to exercise these models may negate the increased accuracy that is possible with more sophisticated models, and the increased costs may not be warranted.

Differences in a program's internal structure can result in significant differences in the estimated heating and cooling load. One should resolve the answers to several important questions before selecting a program:

Is the program model acceptable for the type of building being analyzed and the available data?

Is the program acceptable for the purposes of the modeling?

Is the program cost justifiable, in terms of required and/or expected accuracy?

In summary, one can conclude:

1. There are significant differences among the various programs. These result from user effects (translation of engineering data into the formats required for computer input), different input requirements, and internal differences in program alogithms.

2. Major faults exist in the manner in which these automated techniques compute load components. These faults sometimes cancel each other, leading to total load computations which appear correct.

3. A user who is knowledgeable of buildings and building systems, the basic thermodynamics underlying these models, and the computer systems can validly model the performance of an existing building.

4. Automated techniques can be valid and useful tools in performing sensitivity studies in support of *specific* engineering problems.

5. The use of these techniques for broad policy issues is question-
    able, as is their use in establishing building energy performance
    standards.

This study focused on the load modeling of building energy simulation techniques. With regard to the systems routines, we note
even greater opportunity for internal modeling differences and,
hence, a greater need for rigorous study. This modeling diversity
and its significance are evident to any analyst who has used more
than one system simulation. Finally, we note that the economics
models tend to be tailored to the specific economic measures preferred by the program developer.

NOTES AND REFERENCES

1. ASHRAE simplified method: An assigned percentage of heat gains
    at the hour is treated as cooling load, and the remainder is
    either spread uniformly over an assigned period of time, follow-
    ing the hour or spread according to the "sum of the digits"
    method in decreasing order.

7

Energy Conservation in Residential Buildings

Melvin H. Chiogioji
U.S. Department of Energy
Washington, D.C.

Homeowners are becoming aware of the need to use energy more efficiently. Since the buildings sector accounts for about one-third of
the nation's energy consumption, more efficient energy use in buildings can result in enormous economic savings. But in order to be
able to build better energy-conserving homes, it is important to
understand which combinations of design, operation, and construction
will result in minimum energy use, so that conservation efforts can
be concentrated where they will do the most good. A building consists of three functional elements: the structure, the building
system, and the control system (Fig. 7-1). Each one of these elements has a significant impact on energy consumption, a fact which
must be taken into consideration before a basis for developing energy
conservation techniques can be established.

In considering the energy conservation options available it is
important to choose those which will be of real value in a particular circumstance. The initial cost of a conservation action and the
savings in operating costs which it may provide must be balanced.
To accomplish this balance, the pattern of energy use in a home must
be determined. Table 7-1 gives typical energy use distributions for

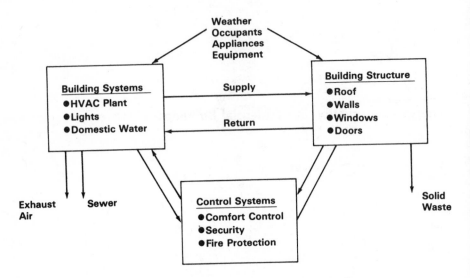

Figure 7-1. Building components functional relationships. (From
*Mixed Strategies for Energy Conservation and Alternative Energy
Utilization in Buildings, Final Report, vol. II, Energy Resource
Center, Honeywell, Minneapolis, Minn., November 1977, p. 18.* Pre-
pared for the U.S. Department of Energy.)

Table 7-1. Residential Energy Use in Midwestern United States

| | South | | Central | | North | |
	On Site	Primary*	On Site	Primary*	On Site	Primary*
Heating	30%	14%	60%	40%	63%	45%
Water Heating	22%	11%	17%	12%	18%	13%
Air Cond.	23%	36%	6%	13%	4%	8%
Lighting & Appliances	25%	39%	17%	35%	15%	34%
Total (MBtu/ft²/year)		156		172		187

*Primary energy consumption includes the generation and transmission
losses associated with electric energy.

Source: Jerold W. Jones, *Energy Conservation Opportunities for Res
Residential Buildings,* The University of Texas, Austin, Tex., updated
1979, p. 697.

a single-family dwelling in three Midwest locations: southern, central, and northern. The house contains 1600 ft^2 of floorspace and
is equipped with both gas space conditioning and water heating and
with typical electrical appliances, including a dishwasher, dryer,
and central air conditioning. In all three cases, the data indicate
that more than half of the energy consumption is related to heating
and cooling.

Energy conservation experts generally agree that in residential
buildings, the greatest energy savings can be attained by reducing
air infiltration through windows and doors and by reducing conduc-
tive heat transfer through the walls and windows. Table 7-2 provides a breakdown of the heating and cooling loads for a characteristic residence. As can be seen, the infiltration load is 55.2% and
41.5% for heating and cooling, respectively. It is also significant
that the internal load created by the use of appliances and lights
and by the occupants is responsible for 35% of the cooling requirements. An analysis of the internal load shows that 25% of the heat
gain comes from the occupants, 19% from the lights, and 56% from
appliances.

Energy-efficient dwellings have been constructed in this
country for a number of years. Many people will remember homes

*Table 7-2. Breakdown of Heating and Cooling Loads for
Characteristic House*

Components	% Of Heating Load	% Of Cooling Load
Ceiling	3.7	2.3
Floor	2.2	2.4
Total Window	13.6	4.1
Total Door	1.4	0.4
Total Wall	23.9	14.2
Infiltration Load	55.2	41.5
Internal Load	—	35.1
	100.0	100.0

Source: Hittman Associates, Inc. *Residential Energy
Consumption, Single Family Housing,* Report No. HUD-HAI-2,
Columbia, Md., March 1973.

designed for cross or updraft ventilation, homes built to take advantage of shade and sun and furnished with small heat sources in
every room, so that the entire house did not have to be heated to
the same thermostatically controlled temperature. Some of us may
even remember the homes built in the early 1900s that were equipped
with solar water heaters. The point is that we should once again
be moving in these directions. An energy-conscious home should use
passive design ideas in its planning, design, construction, and
use. Details such as a building's location and configuration, should
all be carefully considered.

Like the better energy-conserving homes now being constructed
and renovated around the country, the energy-conscious home is well
insulated and incorporates double-glazing and weather stripping
where they are needed--but it should go well beyond that. The remainder of this chapter discusses some of the design features which
can lead to greater energy efficiency in the building structure.
Heating, ventilation, and air conditioning (HVAC) systems, lighting,
and appliances are covered in other chapters.

BUILDING SITING AND ORIENTATION

A building can have maximized energy efficiency by recognizing and
using to advantage the natural elements that surround it. Such a
building will complement its environment and do so in every portion
and operation of the building system. The energy-conscious house
incorporates design ideas in its planning, design, construction,
and use by utilizing solar energy through passive systems. Building location, siting, orientation, configuration, layout, construction, mechanical and electrical system, and interior furnishings
should all be carefully evaluated in terms of their contribution to
energy consumption and conservation. Natural energies should be
used to their fullest.

A building design that ignores the impact of the natural environment will almost always have to use energy in the form of mechanical, structural, or material interventions to compensate for the

resulting discomforts and inconveniences. Clearly, then, a building project should start with a thorough analysis of the assigned site or potential site alternatives. An architect/engineer must understand and anticipate the effects of a particular site or climate on the energy flow of a building if he intends to use the environment advantageously.

Sun

Where energy conservation is a major goal of a building design, the sun is perhaps the single most important natural element to consider. It affects virtually every portion of a building's design: its site and orientation, envelope and glazing, HVAC systems, lighting system, and operating and maintenance policies. In the Northern Hemisphere, for example, the sun factor is most important to the design of the south, east, west, and north sides of a building, in that order. Since the precise effects of the sun vary according to the season of the year and the time of day, accurate calculations of solar loads will be determined by the direction of the sides of a building, the different building height levels, and the time of day.

Controlling solar radiation is one of the best ways the architect/engineer team can reduce energy consumption in a building or group of buildings. Basically, this control consists of maximizing solar energy during periods which require heating (winter) and reducing to a minimum the solar energy that enters the building during cooling periods (summer). Solar heat should not be permitted to enter occupied spaces during the summer; it can, however, be collected with absorption chillers and used to air-condition interior spaces. During the intermediate periods, solar radiation can be collected and stored daily, so that excessive heat received during the day can be released to the interior at night.

For passive solar collection, south-facing glass should total one-quarter to one-fifth of the floor area in temperate climates and one-third to one-quarter of the floor area in colder climates.

These are general rules which, in temperate climates and with appropriate insulated shutters and heat storage mass, can provide
between 40 and 90% of the heating required.

Wind Effects and Ventilation

The velocity and prevailing directions of winds on a given site
should affect the shape and orientation of a building, as well as
the components of its envelope. Wind affects infiltration (air leakage) and transmission (thermal conductance) over the entire envelope
of a building and, in particular, its glazed or windowed portions.
In the Northern Hemisphere, the north and west sides of a building
are most exposed to wind loads.

Winds can decrease the exterior film of still air that usually
surrounds a building and will increase the thermal vulnerability
of roof and wall elements. This can increase heating and cooling
loads. Wind can carry the heat built up by solar radiation away
from a building and can evaporate moisture on wet surfaces, cooling
the building envelope to temperatures lower than ambient air. By
removing water vapors, wind increases humidifying loads. Knowing
the direction of prevailing winds helps to determine where entrances
and exits should be placed and whether or not they should be shielded.
It may also determine the effects of natural versus mechanical ventilation systems and the desirability of one over the other.

Architectural consideration of airflow lies in two major areas:
(1) protecting a building from undesirable winds, and (2) utilizing
the feasible ones for ventilation and cooling.

A building constantly exchanges air with its environment: outside air leaks in, inside air leaks out. A certain amount of this
exchange (approximately one complete air change per hour) is necessary for ventilation, but the exchange in most buildings is
greater than what is needed. In winter, the air that leaks in is
cold, the air that leaks out is warm, and fuel is used to rectify
this temperature difference. This leakage or infiltration is caused
by wind, by the building's acting as a chimney, and by the opening
of outside doors as the occupants enter and exit.

The effects of opening doors and of wind need little explana-
tion, but the chimney effect may not be obvious (see Fig. 7-2). When
air in a building is warmer than the outside air, the entire building acts like a chimney--hot air tends to rise and leak out of cracks
at the upper levels, and this sucks cold air in through cracks at
the lower levels. Both the temperature difference and building
height contribute to this effect. A two-story house with a 68°F
inside temperature and a 30°F outside temperature will produce a
"chimney" leakage equivalent to a 10 m/hr wind blowing against the
building.

Infiltration has traditionally been a major source of thermal
inefficiency in homes. According to various estimates, it accounts
for between 25 and 40% of all heat transfer through the building envelope in both old and new construction. Figure 7-3, developed by
the Texas Power and Light Company, is based on the analyses of 40
single-family homes and shows the aggregate percentage breakdown of
infiltration from various sources.

The absolute magnitude of infiltration losses has declined with
the advent of tighter building components and newer construction
techniques; however, there is a clear potential for large energy

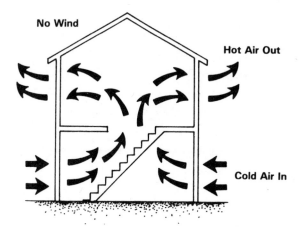

Figure 7-2. Chimney effect.

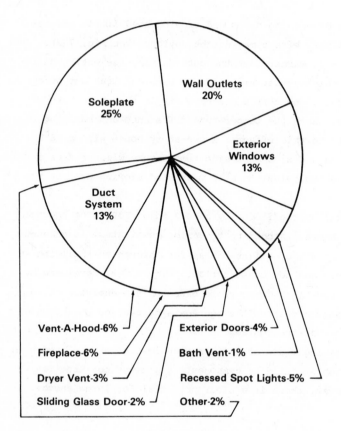

Figure 7-3. Infiltration.

savings through greater infiltration control. The most obvious way
to save the amount of energy required to compensate for infiltration
is to make the building as airtight as possible. But even if this
could be achieved, it would be unwise. Much like a living animal,
a building must breathe in and exhaust fresh air, either naturally
or mechanically. Otherwise, unwanted gases such as radon may concentrate in the building to unsafe levels.

The task of a designer, then, is to protect a building from undesirable winds and to utilize the desirable winds for ventilation
and cooling. One option is to reduce the speed of the wind reaching
the house. Lowering the wind's velocity will reduce infiltration

and, consequently, lower the heating bill. This can be done by placing wind barriers (trees, garage, barn, hedges, or fencing) between
the house and the prevailing winter winds.

If trees or shrubs are used as wind barriers, it is important
to plant vegetation that will result in dense growth and that will
grow to heights equal to or greater than the house. The maximum
distance from the house to the windbreak should not be more than
five times the building height, measured from the leeward wall. An
example of good orientation and shading is shown in Fig. 7-4 (1).
To minimize winter infiltration, evergreen trees should be planted
to the northeast and northwest of the building, so that the cold,
prevailing winter and storm winds do not reach the house. The
double-door entrance should be on the east, away from both the positive and negative pressure sides of the house in winter. For natural summer cooling, a breezeway can be preserved by the planting
arrangement and driveway location.

To summarize, for winter operation air infiltration can be m
minimized by

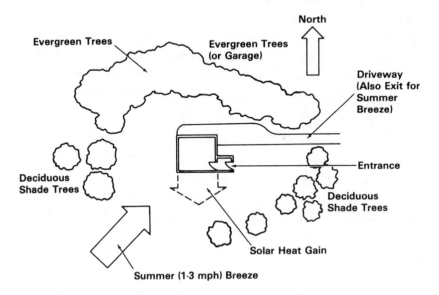

Figure 7-4. Landscaping and orientation.

1. Planting or constructing wind barriers to protect the house from winter storms and prevailing winds.
2. Installing a good vapor barrier which will keep out the cold air and keep in the humidity, ensuring a cozy house all winter.
3. Checking around windows, inside cellar doors, and around ceiling light fixtures for cracks that will permit air to escape or enter and rectifying these faults by caulking and weather stripping.

For summer operation, cool air intake should be maximized by

1. Planting shade trees on the east and west sides of the house. If it is a new house, plan the site with existing trees in mind.
2. Planning to utilize summer breezes (e.g., not building a garage that will prevent the breeze from reaching the house). Also since driveways collect a great deal of heat, avoid placing a hot-top driveway between the house and the summer breeze. However, a pond or shaded area upwind from the home will cool the air before it enters the house.
3. Making maximum use of high and low vents for the removal of hot air and for the intake of cool air.
4. Keeping heat out during the day and introducing cool air at night. For maximum air flow through the house, the outlet vent area should be at least as large as the inlet vent area.
5. Putting turbine vents in the roof and, if needed, installing a two-speed attic fan.

No matter how the house is sited, natural ventilation and cooling can be increased by using casement-type windows or partially opened shutters on the windward side of the building. These projections create minipressure zones in front of the window openings and increase the velocity of the breeze passing into the openings.

Shading

The energy performance of a window can be greatly improved through shading. Internal shading devices, the more common among which are venetian blinds, shades, and draperies, can reject up to 65% of the solar radiation that strikes the glass directly (Fig. 7-5) (2). But since much of the radiant heat that enters the space is at the exterior wall, external shading is most effective against overall heat gain: It can block out up to 95% of the solar radiation that would otherwise enter the building (Fig. 7-6). The general advantage of using exterior appendages to improve window performance is that they mitigate climatic problems before they enter the building. Also, external appendages allow some of the residual forces, such as summer solar heat or winter winds, to be dissipated before encountering the window, although to a lesser extent than can be gained by proper site strategies.

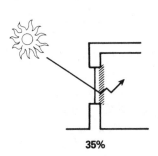

35%

Figure 7-5. Internal shading.

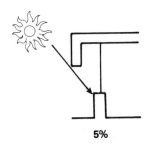

5%

Figure 7-6. External shading.

Many devices are available for exterior shading. The most effective solar control is a simple horizontal overhang along the southern exposure, as illustrated in Fig. 7-7. This blocks the direct sun rays during the summer months, when the sun's angle is highest, but allows the sun's rays to penetrate during the winter, when the angle is lowest. This is a classic principle: Heat is rejected in summer when it is not needed and received in winter when it is most needed.

On the east or west elevations, however, the sun's angle is too low to be blocked out by horizontal overhangs. Properly oriented vertical louvers have proven more beneficial here (Fig. 7-8). If the louvers are movable, the user can control them to provide a better view or a greater diffusion of light at times when the sun is located on the opposite face of the building. If glass must be used on an east or west wall, the low sun angles can be blocked by using a sawtooth wall, as shown in Fig. 7-9. Direct sunlight will be totally eliminated from the building's interior, although both natural daylight and a good view will not.

The effectiveness of a solar screen in shading a window depends on its geometry and on its reflectivity as a material. The geometry determines how high the sun must be above the horizon before the louvers block all the direct sunlight. The reflectivity of the louvers determines how much light penetrates indirectly when it is

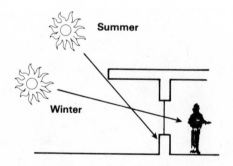

Figure 7-7. Horizontal overhang.

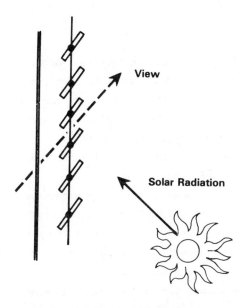

Figure 7-8. Louvered windows.

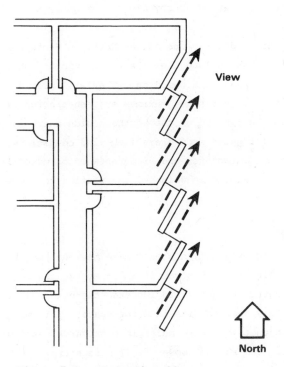

Figure 7-9. Sawtooth walls.

reflected off the surface of the louvers. Also, the color of the projection should be dark to reduce the light reflected off the projection and through the window. The light absorbed by this dark color will be converted to heat and then dissipated to the outside air without becoming an air conditioning load. To ensure this heat dissipation, free circulation of the air is provided by a separating gap between the shading device and the window. Combinations of vertical and horizontal elements can be effectively used to control solar radiation if the proportions of these devices are carefully related to sun angles during the critical times of the day.

Trees can also be effective because they serve as wind and light breaks. They can substantially alter the effects of arctic winds, solar radiation, the flow of air, and winds blowing across arid land or large bodies of water. They affect natural ventilation, air pressure, surface temperatures, and humidity levels, and they shade the paved areas surrounding a building, as well as the building itself.

In the colder regions of the Northern Hemisphere, deciduous trees (trees which lose their leaves in winter) should be planted on the south side of a building. They will provide sunshade during hot months, and yet allow maximum sun penetration in winter. Evergreens can be planted on the northern side where there are no cold-weather solar gains from prevailing wind conditions. In general, the relative advantages and disadvantages of trees in terms of energy conservation should be weighted in relation to the whole building system.

Lighting

The building orientation should make maximum use of sunlight for interior lighting. One way in which energy can be saved by the use of such natural light is to increase the building perimeter and proportionately decrease its interior space. This may result in different adjuncts, such as multiple courtyards, atriums, light walls, skylights, and so on. However, if more energy can be conserved by fewer window areas in conjunction with artificial lighting systems, then a reduction of the perimeter exposure should be considered. Other design considerations include the use of reflective surfaces, such as sloping white ceilings, to enhance the effect of natural lighting and increase the yearly energy saved.

Summary

Figure 7-10 provides a composite schematic site plan which takes advantage of good passive solar designs.

1. First, every effort should be made in the planning, design, and construction of the building to minimize heat loss in the winter and heat gain in the summer.
2. When this objective is achieved, every effort should then be made to incorporate passive design characteristics which use natural energies such as solar radiation, natural light, and prevailing winds.
3. Where possible, economical and efficient, active technologies which use renewable energies should be incorporated.
4. Then, and only then, should fossil-based or nonrenewable energies (and related equipment and design features) be used to supply the remaining energy requirement.

BUILDING ENVELOPES

A building's exterior envelope (its formulation, exterior walls, and roof) is man's first defense against the elements and, as such, is an important target for energy-conscious designs. In this regard, a builder can shape the envelope, turn it, condition it, and seal it according to aesthetic whim, or he can take advantage of its natural surroundings.

The building designed to complement the natural environment may look different on every side because the environment affects it differently on each side. Orientation, shape, exposure, patterns, glazing, thermal wall and roof characteristics, color, texture, and

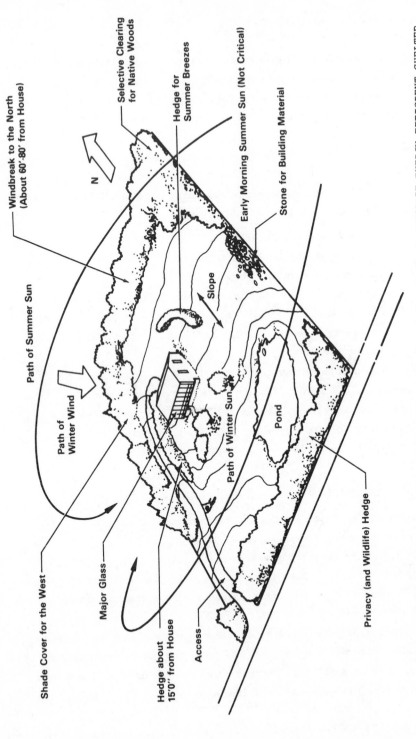

Path of Summer Sun

Windbreak to the North
(About 60'-80' from House)

Selective Clearing
for Native Woods

N

Hedge for
Summer Breezes

Early Morning Summer Sun (Not Critical)

Stone for Building Material

Slope

Path of
Winter Wind

Shade Cover for the West

Path of Winter Sun

Pond

Major Glass

Hedge about
15'0" from House

Access

Privacy (and Wildlife) Hedge

Figure 7-10. Composite of schematic site plan. (Reprinted from Ref. 1, LOWCOST ENERGY EFFICIENT SHELTER © 1976 by Eugene Eccli. Permission granted by Rodale Press, Inc., Emmaus, PA 18049.)

reflective and absorptive surfaces are a few of the relevant considerations. Mechanical systems should supplement envelope design
only when environmental conditions exceed the capacity of the designed envelope to handle them.

Configuration

Residential energy consumption is affected by overall building configuration in various ways. The major variable is the amount of
exposed surface area which, if minimized, will

1. Reduce heat loss and heat gain by conduction, convection, and
   radiation through the building envelope.
2. Reduce heat loss by infiltration through the envelope. Infiltration is a major component of heat loss, and the amount of this
   loss is a function of how much of each room is the exterior wall
   of the building.

To minimize heat transmission for a given enclosed volume, a
building should be constructed with a minimum of exposed surface
area. A round building has less surface area and, hence, less heat
gain or loss than any other shape for an equivalent amount of total
floorspace. A square building has less surface area than a rectangular building of equivalent floorspace, and so experiences less
thermal transmission loss or heat gain. However, the number of
stories modifies this relationship for the building as a whole.

A tall building has a proportionally smaller roof than a short,
flat building and is less affected by solar gains on that surface.
On the other hand, tall buildings are generally subjected to greater
wind velocities, which increase infiltration and heat loss. Tall
buildings are less likely to be shaded or protected from winds by
surrounding buildings and trees, and they require more mechanical
support systems, including longer exhaust duct systems. And since
the chimney effect in tall buildings increases infiltration, special measures are required to reduce its influence on heat gain and
heat loss.

In order to explore the impact of building configurations on
energy demand, a number of representative variations in ceiling
height, location of habitable space, number of stories, and geometric form were analyzed. Each of these variations is briefly described, and their physical characteristics and energy consumption
characteristics are summarized and compared to a standard practice
house (Table 7-3). In each example, the floor area and space complement are equivalent to the standard practice house.

Table 7-4 provides a summary of energy consumption characteristics based on the various configuration options. As can be readily
seen, those configurations that have the least amount of exposed
surface area are the most energy-efficient. Building configuration
is but one of many considerations in energy-conscious design, and
the comparisons in Table 7-3 illustrate the importance of configuration in relation to plan and volume. However, these comparisons
only show the impact that building configuration has on heat loss
and do not reflect any potential benefit or liability caused by
solar gain.

Table 7-3. *Buildings Configurations Options: Physical Characteristics*

Configuration Options	Floor Area S.F.	Volume C.F.	Stories #	Perimeter L.F.	Exposed Walls S.F.	Exposed Ceilings S.F.	Total S.F.
Reduction in Bedroom Height	1,600	12,000	2	114	1,710	800	2,510
One-Story Rectangular (Long)	1,600	12,800	1	178	1,424	1,600	3,024
One-Story Rectangular (Short)	1,600	12,800	1	164	1,312	1,600	2,912
Cube Configuration	1,633	12,800	3	93.3	2,178	544	2,722
Half Dome Configuration	1,623	12,800	2	115	—	—	2,110
3/4 Dome Configuration	1,653	12,800	3	87	—	—	2,413
Square Floor Plan	1,600	12,800	1	160	1,280	1,600	2,880
Circular Floor Plan	1,600	12,800	1	142	1,135	1,600	2,735
STANDARD PRACTICE HOUSE	1,600	12,800	2	114	1,824	800	2,624

Source: Ref. 3, p. 15.

Table 7-4. Buildings Configurations Options: Energy Consumption Characteristics

	HEAT-LOSS (BTUH)	Basement Walls	Basement Floor	Exterior Walls	Windows	Glass Doors	Sliding Doors	Infiltration (CF/HR)	Ceiling	PERCENT HEAT-LOSS SAVINGS OVER STANDARD PRACTICE
		Percentage of Total Heat-Loss								
Bedroom Height Reduction	48,750	6.4	3.3	14.4	27.8	6.4	1.8	33.3	6.6	3
Long Rectangular	49,950	9.7	6.4	11.8	22.4	6.1	1.8	28.8	13.0	1
Short Rectangular	48,280	9.5	6.6	11.2	21.2	6.4	1.9	29.8	13.4	4
Cube Configuration	50,130	5.2	2.2	19.0	27.0	6.2	1.8	34.4	4.2	5
1/2 Dome Configuration	43,208	7.4	4.8	20.5	31.4	7.2	2.1	26.8	—	14
3/4 Dome Configuration	43,166	5.6	2.8	24.3	31.4	7.2	2.1	28.6	—	14
Square Floor Plan	48,000	9.3	6.7	10.9	21.3	6.4	1.9	30.0	13.5	5
Circular Floor Plan	45,660	8.7	7.0	10.1	19.8	6.8	1.9	31.5	14.2	9
Standard Practice House	50,380	6.1	3.2	15.2	26.9	6.2	1.8	34.3	6.3	—

Source: Ref. 3.

Insulation

Heat flows by conduction through building materials and is lost
through the exterior surface of a building. The rate of this loss
as heat is conducted from the warm side to the cold side, depends on
the size of the surface, the duration of the heat flow, the temperature difference between the two sides of the exposed area, and the
type of material used in the construction. While virtually all materials used in building construction reduce the flow of heat, some
materials are more effective in doing so. These are used in insulation.

There are many different types of insulation, both in terms of
form (rigid boards, loose fill, etc.) and of the materials which
comprise them. Table 7-5 gives a comprehensive breakdown of the
variety available. The reflective insulation referenced in the
table are primarily other forms of insulation to which reflective
surfaces have been added, usually as vapor barriers. A vapor barrier is essential for preventing moisture from the inside air, which
can seep through the wall or roof, from reaching the insulation and
other building materials. Without a vapor barrier, the moisture
could rot some types of insulation and, in any case, would inhibit
the insulation's original capability to resist heat flow.

Insulation should be measured·in terms of R (for resistance)
value. The higher a given material's R-value (shown on the package label), the more effectively it resists heat flow. Accordingly,
the amount of insulation needed depends on the amount required to
achieve the R-value desired. The important point, however, is not
a particular type of insulation's R-value; rather, it is the value
needed to reduce heating and cooling energy consumption in a cost-
effective manner.

Table 7-6 lists the insulating value of most of the common
materials found in construction. The R-value shown in the right-
hand column indicates the effectiveness, or resistance value, of
the material. When building sections are made of several materials,
the resistance value of each of the individual materials can be

Table 7-5.   Types of Thermal Building Insulation

Loose Fill Insulations	Fibrous Wool Granular	Rock, Glass, Slag, Wool, Wood Fiber Perlite, Vermiculite, Granulate Cork
Blanket Insulations	1. Plain (No Covering) 2. Open on One Side; Vapor Barrier Paper on Other 3. Enclosed with Paper on One Side and Vapor Barrier Paper on Other 4. Reflective Vapor Barrier on One Side; Other Side Open or Enclosed with Paper	Rock, Glass or Slag Mineral Wool or Wood Fiber or Cotton
Batt Insulations	Same as Blanket Insulations	Same as Blanket Insulations
Insulation Board	Interior Boards Tile, Plank, Sheathing, Roof Insulation, Insulating Roof Deck Shingle Backer, Sound Insulation Board Acoustical Tile	Vegetable Fibers, Mineral Fibers, Plastic Foams
Slab or Block Insulations	Small Rigid Units Usually 1 in. or More Thick	1. Corkboard 2. Wood Fiber and Cement 3. Mineral Wool 4. Insulating Board (Fiberboard) 5. Perlite and Binder 6. Cellular Glass
Reflective Insulations	1. Sheets and Blankets (a) Plain (No Paper Backing) (b) Paper-Backed Foil in Single or Multiple Layers or Accordion Types	Aluminum Foil Plus Other Materials
	2. Aluminum Foil Surfaced Gypsum Board or Other Materials	Aluminum Foil Plus Other Materials
	3. Foil-Surfaced Blanket and Batt Insulations	Aluminum Foil Plus Blanket and Batt Insulations
	4. Reflective Coatings Applied to Paper, Etc.	Coatings Applied to Paper in Single or Multiple Sheets
Plastic Foam Insulation	Available in Slab or Block Forms and Other Types Including Sandwich Panels	Polystyrene, Urethane and Other Types of Plastic

Source: Enviro-Management and Research, Promotional Information for Energy Conservation, U.S. Department of Energy, March 1978, pp. 9-7.

Table 7-6.   Insulation Value of Common Materials

MATERIAL	THICKNESS (Inches)	R VALUE
Air Film and Spaces:		
Air Space, Bounded by Ordinary Materials	3/4 or More	.91
Air Space, Bounded by Aluminum Foil	3/4 or More	2.17
Exterior Surface Resistance	—	.17
Interior Surface Resistance	—	.68
Masonry:		
Sand and Gravel Concrete Block	8	1.11
	12	1.28
Lightweight Concrete Block	8	2.00
	12	2.13
Face Brick	4	.44
Concrete Cast in Place	8	.64
Building Materials—General:		
Wood Sheathing or Subfloor	3/4	1.00
Fiber Board Insulating Sheathing	3/4	2.10
Plywood	5/8	.79
	1/2	.88
	3/8	.47
Bevel-lapped Siding	1/2 × 8	.81
	3/4 × 10	1.05
Vertical Tonge and Groove Board	3/4	1.00
Drop Siding	3/4	.94
Asbestos Board	1/4	.13
3/8" Gypsum Lath and 3/8" Plaster	3/4	.42
Gypsum Board (Sheet Rock)	3/8	.32
Interior Plywood Panel	1/4	.31
Building Paper	—	.06
Vapor Barrier	—	.00
Wood Shingles	—	.87
Asphalt Shingles	—	.44
Linoleum	—	.08
Carpet with Fiber Pad	—	2.08
Hardwood Floor	—	.71
Insulation Materials (Mineral Wool, Glass Wool, Wood Wool):		
Blanket or Batts	1	3.70
	3½	11.00
	6	19.00
Loose Fill	1	3.33
Rigid Insulation Board (Sheathing)	3/4	2.10
Windows and Doors:		
Single Window	—	Approx. 1.00
Double Window	—	Approx. 2.00
Exterior Window	—	Approx. 2.00

Source: ASHRAE Guide and Data Book, 1970 Systems, 345 East
47 St., New York, NY 10017.

added together to obtain the overall total resistance value. This overall R-value can be used to determine the amount of heat loss.

The amount of additional R-value needed depends on many different factors. Generally speaking, the primary determinant is cost. But while more insulation will reduce heat flow, this principle is subject to the "law of diminishing returns." In other words, beyond a certain critical point, the cost of additional insulation will not be worthwhile in light of the increasingly smaller savings it will provide. When the correct amount of insulation is used the savings can be substantial.

For commercial structures, the areas to which insulation might be added are the perimeter walls, the roof, the floor, as well as the ducts and pipes that run through nonconditioned spaces. A thorough inspection should be required, of course, and infrared thermography may be used to indicate the areas that need the most attention.

Since adding insulation to an existing commercial structure is a costly, somewhat complex endeavor, expert assistance should be obtained from local utility companies and from local associations of consulting engineers and insulation contractors. One may also find it helpful to review the contents of ASHRAE 90-75. For information in this regard, contact: American Society of Heating, Refrigerating, and Air Conditioning Engineers, Inc., 345 East 47 St., New York, NY 10017.

The amount of insulation needed depends on the climate and the cost of energy. Tables 7-7 and 7-8 present guidelines which are in line with the Minimum Property Standards developed by the Department of Housing and Urban Development. To calculate the amount of insulation required, first determine from Fig. 7-11 which climate the building is in, then refer to the appropriate column of each table. Note that Table 7-7 is for all types of heating with the exception of electric resistance heat, and Table 7-8 is specifically for electric resistance heat. Table 7-9 provides information on the amount of insulation needed for each R-value shown.

Table 7-7. Insulation Guidelines for Homes Heated with
Oil, Gas, or Heat Pumps

Feature	Zone A	Zone B	Zone C	Zone D	Zone E
1. Ceiling Insulation	R-19	R-19	R-26	R-30	R-38
2. Wall Insulation	R-11	R-11	R-13	R-13	R-19
3. Floor Over Unheated Spaces	none	none	R-11	R-11	R-19
4. Foundation Walls of Heated Spaces	none	none	R-6	R-11	R-11
5. Slab Foundation Perimeter	none	R-2	R-5	R-5	R-7.5

Source: Tips for Energy Savers, U.S. Department of
Energy, 1980.

Table 7-8. Insulation Guidelines for Homes Heated
with Electric Resistance Heat*

Feature	Zone A	Zone B	Zone C	Zone D	Zone E
1. Ceiling Insulation	R-19	R-22	R-30	R-30	R-38
2. Wall Insulation	R-11	R-13	R-19	R-19	R-19
3. Floors Over Unheated Spaces	none	R-11	R-19	R-19	R-19
4. Foundation Walls of Heated Spaces	none	none	R-6	R-11	R-11
5. Slab Foundation Perimeter	none	R-5	R-7.5	R-7.5	R-7.5

*Based on the proposed revisions to the HUD Minimum
Property Standards, April 1978.

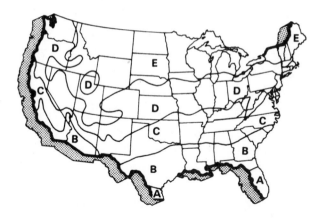

Figure 7-11. Climate zones of the United States.

Table 7-9. R-Values Chart

	Batts or Blankets		Loose Fill (Poured In)		
	Glass Fiber	Rock Wool	Glass Fiber	Rock Wool	Cellulosic Fiber
R-11	3½"-4"	3"	5"	4"	3"
R-13	4"	4½"	6"	4½"	3½"
R-19	6"-6½"	5¼"	8"-9"	6"-7"	5"
R-22	6½"	6"	10"	7"-8"	6"
R-26	8"	8½"	12"	9"	7"-7½"
R-30	9½"-10½"	9"	13"-14"	10"-11"	8"
R-33	11"	10"	15"	11"-12"	9"
R-38	12"-13"	10½"	17"-18"	13"-14"	10"-11"

Source: Tips for Energy Savers, U.S. Department
of Energy, 1980.

Windows

Windows serve many functions in buildings: they allow daylight to
enter, provide a view of the outside with its changing weather patterns, provide ventilation, and admit heating in the form of sunlight. They also allow heat to escape the building, both by infiltration around the frame and by the normal radiative-conductive-convec-
tive heat transfer process.

From the standpoint of energy-efficient building operation, the ideal window would allow, with very low thermal loss, both the visible sunlight and the invisible infrared rays that comprise over half of the sunlight to enter the building whenever heating is needed. During the summer, when no heating is needed, while providing a view of the outside, it would admit only visible light, and only in the quantities needed for lighting a room.

A number of basic design and operational factors are critical to a window's performance. These include: size of the window, its orientation and exterior shading, its ability for thermal resistance, reduction of air leakage, the use of internal and external window coverings, and use of daylight. In the last several years, increasing attention has been devoted to new window products that will add to the flexibility of window use and substantially improve its overall energy efficiencies. The successful implementation of these products and designs should make it possible for windows to lower the overall energy requirements for space conditioning. For example, while the basic methods for conduction and convection control (in addition to multiple glazing) have been the use of various forms of blinds, shutters, and drapes, and while some sunscreen devices can also baffle the outer air layer to reduce surface conduction, relatively little attention has been paid in existing windows to the control of convection between glazed layers. One technique that has been studied is that of filling the interglazing space with heavier molecular weight gases, such as carbon dioxide. Heat loss can also be reduced by attaching a piece of aluminum foil directly to the inner surface of a window.

The energy efficiency of windows can be increased through the proper use of conventional materials and components. In many cases, new technologies are improvements over the present underutilized state of the art. However, none of these new technologies can be labeled the single "best" approach for all conditions: Each differs radically from the other in cost, effectiveness, type of effect, and range of applicability. Some of the salient features of these new technologies are pointed out in Table 7-10, which also presents the

performance parameters, operating requirements, estimated cost-
effectiveness, applicability to retrofit, and current status of
each new technology. The estimates of cost-effectiveness are qualitative and approximate.

Where energy costs can be reduced with little or no additional
investment cost, as by changing a building's orientation from north
to south, the least-cost energy decision is also the lower life-
cycle cost decision. However, when capital and labor costs are increased by choosing one window design or accessory over another,
these cost increases must be weighed against energy cost savings to
determine the most cost-effective decision.

In summary, it is suggested that the following items be considered in the effort to promote the construction and operation of
more energy-efficient and cost-effective homes:

1. Orient the window away from the north and away from prevailing
   winds in order to cut energy costs substantially.
2. Encourage the use of double-glazing or storm windows in moderate
   to cold climates, particularly on north walls and for large windows.
3. Encourage the use of designs which provide good natural lighting, particularly in rooms with predominantly daytime use.
4. Encourage the use of window coverings, such as draperies, blinds,
   shades, or shutters, which have good thermal resistance and
   shading potential and which complement daylight utilization.

Roof Design

The primary way to enhance the energy efficiency of roofs has been
to install ceiling insulation, but the number of new ways in which
roofs can be designed can also contribute to energy efficiency (3).
Double-roof construction is familiar to anyone who has seen an old-
fashioned icehouse--a partially underground structure whose speci-
ally designed roof allows one to store ice harvested in the winter
for summertime use. The icehouse roof is comprised of an outer roof
of shingles and sheathing on wood nailers, which in turn are secured

Table 7-10. Summary of New Fenestration Technologies

Name of Technology	Winter		Summer		Effect on Lighting/Outlook When Operating	Operability	Estimated Cost-Effectiveness	Applicability to Retrofit	Current Status
	Radiative Gain	Heat Loss (All Types)	Radiative Gain	Other Gains					
1. Double-sided blind	≡90% of unshaded u.v. is controlled	moderate reduction	moderate reduction	moderate reduction	no outlook; very low light	manual; seasonal reversal of blind required	very high	very high	ready for commercialization
2. Triple blind	≡90% of un shaded u.v. is controlled	moderate-high reduction	moderate reduction	moderate reduction	no outlook; very low light	manual, somewhat complex	high	high	on market
3. Shades between glazing with heat recovery	some loss but distribution/storage capability; u.v. is controlled	moderate reduction	moderate-high reduction	moderate reduction	some outlook; lighting possible when in operation	manual	high	low	ready commercialization
4. Between glazing convection and radiation control	≡90% of unshaded; slats can direct Sun away from furnishings	high reduction	moderate-high reduction	high reduction	no outlook or light; some degradation of outlook at all times	none	moderate-high	low	ready R&D
5. Insulating shades and shutters	N/A	very high reduction	high reduction	very high reduction if closed	no outlook; no light	manual; or automatic	low-moderate	low-moderate	products marketed
6. Skylid®	N/A	high reduction	high reduction if closed	high reduction	no outlook; no light	automatic	low-moderate	low	on market

7. Beadwall®	N/A	very high reduction	high reduction if filled	very high reduction if filled	no outlook; no light	automatic	low moderate	low	on market
8. Beam daylighting	beneficial distribution effects and u.v. control	N/A	no effect	reduced lighting load	increased daylighting; affects upper part of window only	manual; only seasonal adjustment is essential	low-moderate	low	advanced R&D
9. Selective solar control reflective film	significantly reduced	N/A	moderate-high reduction	N/A	none	none	high in pre-dom. cooling climates	very high	R&D
10. Heat mirror	low reduction	low-moderate reduction	low reduction	some reduction	none	none	moderate-high	very high	on market retrofit package near marketing
11. Optical shutter	low reduction	N/A	high reduction	N/A	no outlook when translucent	automatic	probably low	probably low	R&D
12. Weather® panel	low reduction unless overheating occurs	very high reduction	high reduction	very high reduction	no outlook	passive/automatic			R&D
13. Optimization of fenestration design (size, orientation)	significant optimization over current practice	N/A	high reduction through proper shading, etc.	reduction	N/A	N/A	very high	very low	early R&D

Source: Office of Technology Assessment, *Residential Energy Conservation*, U.S. Congress, July 1979, p. 239.

to an inner roof of roofing felt and sheathing on rafters. The 3-
to 4-in. airspaces between the nailers are left open at the eaves.
In the summer, the outer roof heats up, but through convection, the
ventilated air space between the two roofs carries away much of the
heat. Figure 7-12, based on an icehouse-type of roof replacing a
conventional roof, demonstrates the impact of this concept. The
outer roof has a U-value of 0.83, and with R-33 insulation the inner
roof has U-values of 0.031 (summer) and 0.1030 (winter).

Other roof designs which show promise are modulated roof systems. For example, an operable skylight developed by Zomeworks,
Inc. allows penetration of light and heat when desired and automatically closes when conditions require it (see Fig. 7-13). Canis-
ters of freon gas are used to provide energy control.

Another approach to the modulated roof has been patented by
Harold Hay of Sky Therm Process and Engineering, Los Angeles, California (see Fig. 7-14). In this system, a flat roof structure sup-
ports black plastic bags, each of which is 8 in. deep and contains

Figure 7-12. Icehouse roof. (From Ref. 3.)

Figure 7-13. Skylids. (From Ref. 3, Zomeworks, Albuquerque, N. Mex., Patent no. 3884414.)

noncirculating water. During the day, the bags are exposed to the sun and act as both heat storage devices and radiators to space below. At night, insulated shutters on the roof side are closed to keep heat from escaping to the night air. In warm weather, these shutters remain closed during the day and are opened at night. Hay's 36 × 52 ft structure uses 7000 gal of water, providing the heat storage capacity of 16 in. of concrete at a deadweight equivalent to only 4 in. of concrete.

Other passive approaches to energy-conscious roof design include roof ventilators, rotating skylights, and water springs.

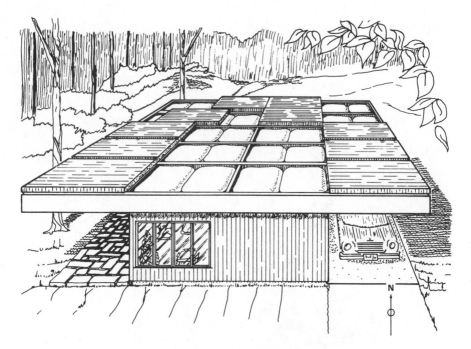

Figure 7-14. Sky Therm roof. (From Ref. 3, patented by Harold Hay, Sky Therm Process and Engineering, Los Angeles, Calif.)

Greenhouses and Atriums

Windows can account for between 15 and 30% of the total heating energy loss in a house. Therefore, placing them strategically can make significant differences in both reducing the heat loss and maximizing solar gain in the winter. One approach to achieving these objectives is to face windows onto an interior atrium. An atrium is an open court, centrally located within a structure, sur-rounded by enclosed space and open to the sky. If a major fraction of a building's exterior glass area can be oriented inward to an atrium, substantial energy savings will result (1) by reducing heat losses through and around windows on the exterior wall of the structure, and (2) by using the atrium as a passive solar collector.

Atriums can also serve as positive design features: They can open up the floor plans, provide visual privacy from unit to unit, and, where desired, allow higher densities. To maximize the passive solar collector aspect, the atrium should be covered with a sky-light and an insulating shutter to keep the heat in when the sun goes down and to provide shading during times of unwanted solar gain.

To further reduce energy losses through walls and windows, a greenhouse can be placed against one of the outside walls of the house. The greenhouse effect is well understood as a principle of solar collection: The greenhouse reduces infiltration loss by serving as a second wall assembly outside the regular wall and it reduces heat loss by effectively raising the temperature just outside the regular wall (a result of solar gain in the greenhouse).

The intricacies of greenhouse design as they relate to plant growth and the environmental controls necessary to sustain growth are beyond the scope of this chapter. For passive design purposes, however, these simple rules of thumb may be of value:

1. A greenhouse can be constructed as a secondary skin on any facade, but a south-facing greenhouse makes the best passive solar collector.

2. The wall between the greenhouse and the residence should be insulated to control heat flow and to prevent overheating of the house.

3. Wall surfaces directly exposed to the sun should be dark-colored. Floors should be light-colored if there is danger of them becoming too warm.

4. Insulated shutters allow the greenhouse to retain the heat collected after the sun has gone down.

5. The distance from the exterior wall of the greenhouse to another object on the landscape (a tree, another building, etc.) should be at least $2\frac{1}{2}$ times the height of the wall.

ENERGY SAVINGS POTENTIAL

Energy savings potential for diverse improvements made to buildings differ depending on the type of building, its mechanical systems, design, geographical location, the type of fuel used, and occupant practices. In a study done for the Department of Housing and Urban Development (HUD), Hittmann Associates surveyed and evaluated characteristic building designs and practices in 11 selected metropolitan areas (4). Table 7-11 presents their analysis of the annual energy use data in these cities.

As is readily evident, energy usage varies considerably from region to region. Given any energy type, the most important factor in determining residential energy use is climate. Buildings in cold climates such as Boston, Chicago, Denver, and Minneapolis will require large amounts of energy for heating. Buildings in hot climates such as Houston and Miami will demand energy for cooling. Thus, since no existing technology can significantly modify the weather, building characteristics are the most important factors in determining energy use. It also follows that designing buildings below reasonable levels of thermal integrity can result in extraordinarily large energy requirements.

Table 7-12 presents structural and system improvements selected for each building in the Hittman study. The following three-step process was used to assure that the most effective combination of improvements would be selected.

1. The heating and cooling loads were compared to determine whether either one was the dominant factor in the building's total energy use.

2. Based on the above comparison, the component parts of the appropriate load were examined to determine where improvements would be most effective in reducing the building's energy requirements.

3. Modifications to improve the deficiencies of the characteristic buildings were selected and then checked against the original restrictions to make sure that these changes were technically feasible, currently available and so forth.

The HVAC systems were evaluated independently of the structures and were either modified to increase their performance or replaced

Table 7-11. Characteristic Single-Family Residences Annual Energy Data

	Atlanta	Baltimore	Boston	Chicago	Denver	Houston	Los Angeles	Miami	Minneapolis	San Francisco	St. Louis
Heating Load Per Unit (Therms)	352	710	915	998	1478	253	255	86	1411	454	689
Cooling Load Per Unit (Therms)	574	282	90	178	361	835	360	1326	91	70	219
In-Structure Heating Energy Use Per Unit (Therms)	490	1010	1285	1403	2065	356	345	90	1996	608	975
In-Structure Cooling Energy Use Per Unit (Therms)	325	125	48	96	200	467	185	774	53	30	117
Total In-Structure Energy Per Square Foot (Therms/Sq. Ft.)	0.48	0.76	1.11	0.88	1.22	0.48	0.31	0.51	1.40	0.44	0.96
Primary Heating Energy Per Unit (Therms)	506	1044	1328	1447	2130	367	355	295	2054	627	1005
Primary Cooling Energy Per Unit (Therms)	1053	388	152	309	648	1516	606	2492	170	99	384
Total Primary Energy Per Square Foot (Therms/Sq. Ft.)	0.92	0.96	1.23	1.03	1.50	1.10	0.56	1.63	1.52	0.50	1.22

Source: Ref. 4.

Table 7-12. Modifications to Single-Family Buildings

Modifications Structural:	Atlanta	Baltimore	Boston	Chicago	Denver	Houston	Los Angeles	Miami	Minneapolis	San Francisco	St. Louis
Glass Reduction on North Face[1] (%)	25	25	50	44	50	25	25	25	50	25	25
Glass Reduction of South Face[1] (%)	25	25	7	0	5	25	25	25	9	25	25
Addition of Weatherstripping	•		•	•	•	•	•	•	•	•	•
Use of Storm Windows or Double Glazing		•	Existed	Existed					Existed		
Use of Reflective Glass						•		•			
Shading Southern Building Face (Seasons Noted)	Summer					All Year		All Year			Summer
Addition of Wall Insulation Up to Indicated R Value	17	11	17	17	17	17	17	11	17	17	17
Addition of Ceiling Insulation Up to Indicated R Value	27	16	27	27	27	27	27	27	27	27	27
Addition of Floor/Perimeter Insulation Up to Indicated R Value	10	4	12.5	12.5	12.5	10 Existed	10 Existed	10	12.5	12.5	12.5
System.											
Improved Furnace/Heat Recovery System	•		•	•	•	•	•	•	•	•	•
Substitution of Heat Pump for Electric Resistant Heating								•		•	
Use of Improved Cooling System	•		•	•	•	•	•	•	•	•	•

Source: Ref. 4.

entirely with more energy-efficient systems. No changes in energy type were made.

After the improvements presented in Table 7-12 were incorporated into the characteristic buildings, hourly heating and cooling loads were computed. Figures 7-15 and 7-16 illustrate the improved buildings' annual total primary energy use on a per-living-unit and per-square-foot of floor area basis. One conclusion which can be

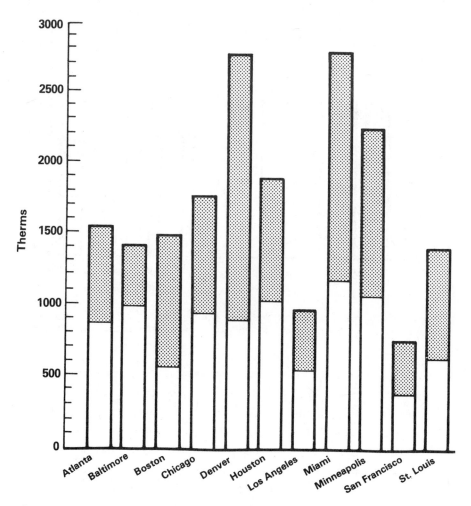

Figure 7-15. Total annual primary energy use per unit for characteristic and improved residences. (From Ref. 4.)

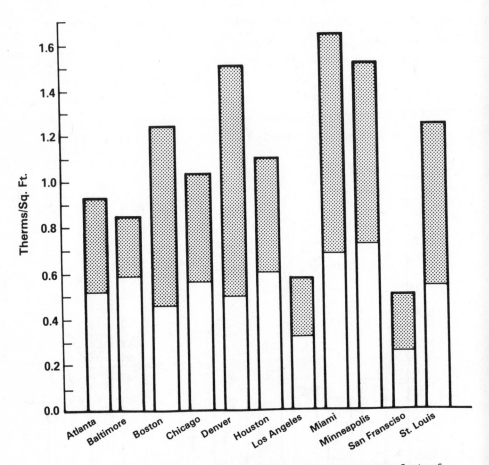

Figure 7-16. Total annual primary energy use per square foot of floor area for characteristic and improved residences. (From Ref. 4.)

drawn from these figures is that current building practices and designs are not as energy-efficient as they could be. Several cost-
effective modifications can be identified, and energy use reductions on the order of 50% are possible.

The single-family residences proved to be the most energy-intensive buildings and exhibited the largest energy savings potential.
Townhouses were next in both categories, and low-rises were the
least energy-intensive. High-rises showed the smallest potential

savings. The choice of heating energy is the single most important factor in determining total primary energy use, followed by the climatic environment (with hot regions being more demanding than cold regions) and by the thermal integrity of the buildings.

CASE STUDIES

As has been shown, energy conservation can now reap significant and continuous dividends. Energy-conscious designs which can provide energy savings in homes begin with choosing the right site and properly locating the house to take advantage of the sun, wind, and other natural forces. The house's configuration, exterior fixtures, and interior characteristics all contribute to its energy efficiency. The following three studies show what can be done to enhance energy efficiency.

The Minimum Energy Dwelling (5)

The minimum energy dwelling (MED) research project was conceived in 1975 by the Southern California Gas Company. The primary purpose of the project was to investigate the energy conservation techniques, building materials, and current technologies available to today's home builders, and to assess the influence of occupant lifestyles on residential energy consumption.

In order to execute and administer the project, the Southern California Gas Company entered into a joint venture with the Mission Viejo Company of Mission Viejo, California, and the Department of Energy. As a progressive residential building developer, the Mission Viejo company was interested in exploring ways of conserving energy in the homes they built. The company was responsible for the design and construction of the two 1150-ft^2 detached single-family homes (Cordova model), one of which is a model for the MED demonstration. It was left unoccupied to provide a base against which the impact of the residents in the remaining MED home could be compared. The performance of the MED homes was also compared with data acquired from a similar, conventional or "non-MED" home in the same area.

Although many of the goals established in the MED program are
project-specific, the majority of the energy conservation techniques, if suitably modified, are applicable to residential construction elsewhere in the country. MED uses currently available equipment which is either cost-competitive or sensible on a life-cycle
cost basis. The one exception is the solar-gas energy system, which
is more expensive than conventional equipment but which can save on
natural gas consumption by using solar radiation as a primary energy
source, with gas as a backup. It is expected that solar energy
equipment will become increasingly cost-competitive with conventional
energy systems in the future.

The final design of the MED homes contained many ideas which
contribute to energy savings (see Fig. 7-17).

1. Exterior walls are built of 2 × 6 in. studs, spaced 24 in. from
   center to center. This practice allows for more insulation and
   a stronger structure, and it uses the same amount of wood found
   in typical home construction (2 × 4 in. studs on 16-in. centers).
   Rafters are built with 2 × 10 in. timbers.

2. The slab floor and the 2-ft underground footings under the exterior walls are poured with 2 in. of insulation laid along the
   inside edge of the footing and under the slab. Slab insulation
   extends 2 ft into the interior living space.

3. A putty-like sealer (mastic) is placed between the sill plate
   and concrete, in order to reduce outside air infiltration.

4. Areas which are difficult to insulate after the exterior walls
   have been closed in, such as corners and around windows, are
   insulated early in the construction process.

5. The roof extension (eave) is lengthened for better shading over
   windows and walls.

6. Vertical shading "wings" protrude from the exterior walls on
   each side of the windows to reduce the amount of sun hitting the
   windows. Double-pane windows, with adjustable shutters between
   the panes, allow the sun to be blocked out.

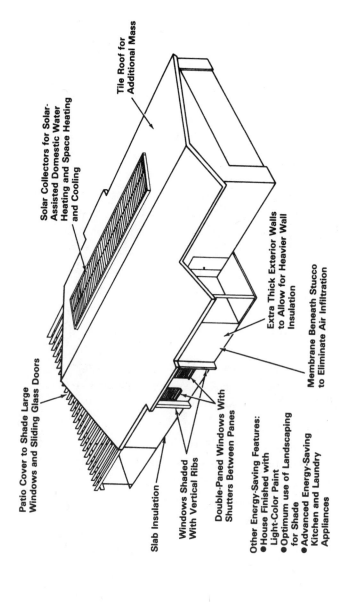

Tile Roof for Additional Mass

Solar Collectors for Solar-Assisted Domestic Water Heating and Space Heating and Cooling

Extra Thick Exterior Walls to Allow for Heavier Wall Insulation

Membrane Beneath Stucco to Eliminate Air Infiltration

Patio Cover to Shade Large Windows and Sliding Glass Doors

Slab Insulation

Windows Shaded With Vertical Ribs

Double-Paned Windows With Shutters Between Panes

Other Energy-Saving Features:
● House Finished with Light-Color Paint
● Optimum use of Landscaping for Shade
● Advanced Energy-Saving Kitchen and Laundry Appliances

Figure 7-17. Minimum energy dwelling. (From Ref. 5.)

7. Patio covers are used to shade the large sliding glass doors
    at the rear of the home. A shaded entryway protects the front
    windows.

8. Under the stucco exterior, plastic, rather than standard build-
    ing paper, is wrapped around the wood structure. This serves
    as another positive barrier against outside air infiltration
    and moisture.

9. Insulation is thicker than normal for homes in the vicinity,
    with higher resistance value in walls (R-19) and attic (R-30).

10. The front door is foam-filled steel with magnetic weather strip-
    ping. This reduces air infiltration and is an improvement over
    the standard door insulation used in typical construction. In
    addition, double-entry doors are used--one leading to an entry
    area, and a second leading into the living room.

11. The domestic water system, which supplies the kitchen and bath-
    rooms, uses one pipe which blends hot and cold water at the
    water heater. It then distributes the desired water tempera-
    ture for a particular use at a particular outlet in the home.

12. Energy-saving, water-restrictive devices are used on all domes-
    tic water outlets in the home.

13. Red tile is used on the roof to lengthen the time it normally
    takes to heat up the home's living space on warm days.

14. Household appliances have energy-conserving features, such as
    a pilotless ignition system on the kitchen range, heat-pipe
    water heaters, and a night setback thermostat on the furnace.
    Heat rejected from the refrigerator compressor and condenser
    coil is channeled outside during the summer and into the home
    during the winter.

15. The landscaping and exterior paint are all planned to help maxi-
    mize energy conservation.

The MED homes are heated and cooled with solar energy, and the
domestic hot water is preheated with solar energy (see Fig. 7-18).
The comfort control systems use natural gas as the backup fuel for
the solar-generated energy supply. Natural gas is used only when
the solar energy is insufficient.

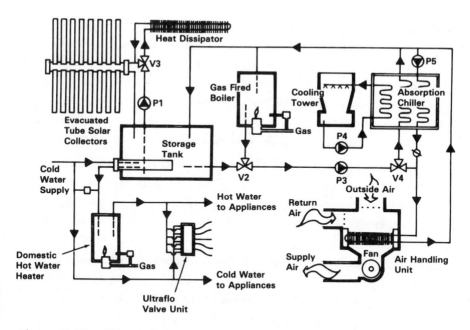

Figure 7-18.   The comfort control system.   (From Ref. 5.)

The project has been a success. The MED homes exceeded their
goal of reducing net energy consumption by 50%. This was the collective result of an extremely tight, well-insulated structure, the
sophisticated mechanical system, and the use of energy-conserving
appliances. Many of the energy-conserving ideas used in the MED
homes have proven cost-effective and desirable from either a construction or marketing viewpoint. As a matter of fact, a number of
the more typical conservation items used in the MED homes have already gained widespread acceptance as salable features in the building industry. These include deeper wall and roof framing members
with increased insulation throughout, more effective vapor barriers,
insulated steel entrance doors, quality windows, and weather stripping.

The project has also spurred interest in and subsequent research on other MED conservation techniques. The most promising of
these seems to be the use of outside air for economizer cooling.
The MED homes use the principles of building mass to delay the mid-

afternoon peak cooling demand to the evening hours, when ambient conditions permit the use of outside air to cool the structure.

One of the most valuable aspects of the MED project was the opportunity to confirm initial design calculations and intuitive assumptions formulated in the early stages of the project. These experiments resulted in test information that was both expected and surprising. Some of the more significant results are as follows:

1. The overall thermal resistance of the MED homes was found to be R = 16.2 compared to R = 7.6 for the standard Cordova. In fact, the measurements indicated a slightly higher R-value than what was calculated, but the test results are within the realm of experimental uncertainty.

2. Methane decay infiltration tests conducted by the Honeywell Corporation indicated standing infiltration rates of 0.25 air changes per hour, or an average of 41 cfm, in the MED units. This is approximately one-fifth that of conventional construction and is indicative of an extremely tight structure. This suggests that a minimum infiltration structure can be produced through quality construction and currently available materials.

3. The mechanical system as a whole functioned as predicted. It should be noted that there was comparatively little trouble with the control logic and subsequent system operation, considering the experimental status of the collectors and the sophisticated nature of the components. This suggests that solar-gas space conditioning systems of this complexity are technically feasible from an operations standpoint. However, there is still a great deal of work to be done to optimize mechanical equipment for low tonnage cooling and small heating demands in energy-conscious dwellings of the MED caliber.

From its conception, the MED project has been a highly valuable educational resource. One of the most heavily instrumented and monitored projects of its kind, the homes have supplied continuous data on a multitude of building envelope and mechanical system functions. Of equal importance is the degree of public awareness which has resulted from the MED project and the quality of information which has been made available to the public.

Brownell Low-Energy Requirement House (6)

In 1977 Bruce Brownell designed and constructed a 2300-ft^2 traditional New England "saltbox" house. The house incorporates in one structure most of the key design elements he has worked with over the past 15 years: (1) the use of superior insulation (urethane board stock) combined with infiltration and vapor barriers, not only on the exterior walls and roof but below the house as well; (2) the application of insulation on the outside, rather than the inside, of the building envelope; (3) the provision of a significant area of south-facing glass for passive solar gain; (4) the use of substantial comfort range thermal storage units, principally in the form of a large sand mass which contains air ducts and which is located below the cellar floor (but within the insulation envelope); and (5) the use of low-speed fans for continuous circulation of air through the occupied space and around the thermal storage mass. It was intended that the heating requirements of this house, in the 8300 degree-day Adirondack environment, would be met almost entirely by heat from three sources: (1) waste heat from appliances (2) passive solar gain, and (3) metabolic heat from the occupants. Only a small amount of supplemental electrical resistance baseboard heating or of heat from a wood stove was expected to be required in the most severe winter weather.

The house employs the post and beam construction common in the Adirondack area. It has 387 ft^2 of glass, of which 250 ft^2 (65%) is south facing, 57 ft^2 (15%) east facing, 72 ft^2 (18%) west facing, and only 8 ft^2 (2%) north facing. The north glazing is located in the kitchen. The house has 150 tons of thermal storage consisting of 2 ft of sand (115 tons) below a concrete floor, concrete cellar walls, and a large 20-ton chimney-like thermal mass rising in the building's center. The central mass contains downflow circulation air ducts on either side of the smoke flue; these air ducts connect to the thermal storage mass below the floor (see Figs. 7-19 and 7-20).

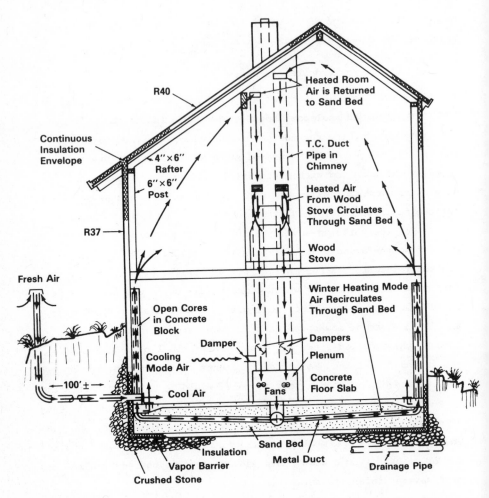

Figure 7-19. Diagrammatic section of LER house. (From Ref. 6.)

The basic roof lines are stylistically New England saltbox, although neither the 250 ft^2 of south-facing glass nor the north elevation with its single window are typical. The house is situated
on a south slope (45° north latitude), with two-thirds of the lower
level of the southern elevation accessible to grade. The grade is
leveled to the south to provide a reflective (snow-covered) surface
for additional solar gain. Brownell has plans to provide optional

Figure 7-20.   Cross-section view.   (From Ref. 6.)

movable insulated shutters for the windows, which could provide
1000 Btu/hr savings at night.

A key feature of the house is an envelope designed for superior
thermal performance. As a first step in the thermal analysis, a
steady-state heat loss calculation was performed for the house. For
purposes of comparison, similar calculations were performed for
three other envelope designs of houses with the same amount of floor
space and similar exterior appearance. These calculations were based
on a 68°F inside temperature and a 0°F outside temperature. An air
change rate of 0.5/hr was assumed for the Brownell house (occupied
mode). Higher rates of 0.75 were assumed for the Arkansas and HUD
designs and 1.0 was used for the inventory house. The three comparison houses were assumed to have fiberglass batt insulation, and
the manufacturer-specified R-values were used in the calculations.
The results are shown in Table 7-13.

Table 7-13. Calculated Building Envelope Performance ($0^{\circ}F$)

Design	Heat Loss, Btu/Hour	Heat Loss, Btu/Day	Normalized Heating Requirement (NHR), Btu/$^{\circ}$F-Hour
Brownell LER House	38,171	916,100	561
House of Arkansas-Type Construction	53,533	1,284,800	787
House Designed to HUD Minimum Property Standards	62,623	1,503,000	921
Typical Inventory House*	93,419	2,242,000	1,374

*A representative 15- to 20-year-old house in the northeast United States.

Source: Ref. 6.

The 38,171 Btu/hr calculated heat loss of the Brownell house is 29% less than that of the Arkansas-type house (7), 39% less than that of a HUD minimum property standards house, and 59% less than that of a typical house in the northeastern U.S. inventory. Because of the high thermal performance of Brownell's wall and roof construction, the heat lost through glazing represents a large fraction, approximately 39% of the total envelope heat loss. Shuttering the windows at night by using a material with an R-value of 5 would reduce the heat loss by 30% to about 26,6000 Btu/hr on a 0°F day.

Table 7-14 provides a summary of the estimated heating requirements for each house for the New York City area. It is evident that dramatic reductions in annual fuel usage can result from a combination of improvements in the building envelope and utilization of modest amounts of internal heat sources and solar gain, provided that adequate diurnal storage capability exists. As indicated in Table 7-14, the Brownell house would probably use only about 22% of the oil needed to heat a typical inventory house in the Northeast. But the low break-even temperature of houses like Brownell's can

Table 7-14. Estimated Annual Fuel Usage for Space Heating

	2,300 Ft.^{2} Brownell House	2,300 Ft.^{2} "Arkansas" Type House	2,300 Ft.^{2} HUD Minimum Property Standards House	2,300 Ft.^{2} Inventory Type House
NHR, Btu/°F-Hr.	561	797	921	1,374
Break-Even Temp., °F	50.8	55.7	57.5	61.0
Degree Hrs. Below Break-Even Temp.[1]	56,158	77,479	86,929	102,182
Annual Heating Load, Btu/Yr.	31.5×10^6	61.5×10^6	80.1×10^6	140.4×10^6
Estimated Annual Fuel Oil Usage, Gals.[2]	388	757	987	1,729
Annual Fuel Cost @ 55¢/Gal., $	213	416	543	951

[1]Based on 1-year hourly weather data for New York City area.

[2]An oil-fired heating system is assumed with a seasonal efficiency (n_s) of 0.58.

Source: Ref. 6.

cause problems outside the heating season. The winter benefits of internal heat sources and solar gain can become liabilities when the ambient temperature exceeds the break-even temperature (51°F for the Brownell house) by a significant amount. The result would be overheating of the occupied space and, possibly, a need for cooling. However, with careful application of design features such as blinds, shades, and architectural roof overhangs, the effects of direct solar gain can be reduced (8). The remaining solar gain and internal source load can be dealt with through an economizer cooling mode (controlled ventilation) which uses outside air. This would be effective at outside temperatures up to about 78°F. Above that temperature, some mechanical cooling would be required. The number of cooling degree hours at above 78°F is usually quite small in comparison with heating degree-hours in regions where heating requirements are high. The cost for cooling, if any, would be relatively small.

The Brownell design concept offers a simple energy system which can be easily understood and which requires a minimum of maintenance. These two features should lead to a general preference of this design over more complicated energy systems with unknown service and performance lives. The house does not require space age technology; it is simply constructed of readily available materials of known life and performance.

The amount of fuel savings possible through efficient thermal envelope design alone is impressive. With the addition of thermal storage concepts, the utilization of "free heat" becomes feasible, and the energy conservation potential of a total concept like Brownell's can be realized. The construction on a large scale of low thermal-loss houses which have a reasonable capacity for thermal storage would not only save a great deal of energy but would, in effect, give utilities serving electric heat customers a remote storage network which could mitigate the utilities' peak loads.

Brownell calculates the possible "free heat" gain in his salt-box house (family of four) as follows: 682 Btu/house for metabolic load, 4677 Btu/hr living cycle heat, and 6726 Btu/hr solar gain, for a total of 12,805 Btu/hr. (An additional 1000 Btu/hr would be available if insulating shutters were provided.) This free heat, together with the energy conserved through the improved thermal integrity of the building envelope, results in a substantial reduction of heat required from the house's auxiliary heating system on a winter day. The optional wood stove used by Brownell delivers about 30,000 Btu/hr. He calculates that the stove would provide the balance of the required heat under winter conditions even if it were operated as little as 12 to 15 hr/week. Thus, in areas where wood is readily available, the house has the potential of requiring only small amounts of fossil or electric energy for its space heating.

According to Brownell's estimates, the cost of building the house is about $9250, or $4.02 more per square foot than the cost of

conventionally constructed houses of similar size. His detailed
estimate of the cost differences is provided in Table 7-15. He
estimates that the purchased energy needed to heat the house is
about $750 to $850 less per year than that required for a conventional 2300-ft^2 house in his area of the country. On that basis,
he suggests that the energy-conserving features of the house have
a simple pay-back period of about 11 to 12 years. He feels that
the period can be shortened to 8 to 10 years if probable future
energy price increases are taken into account.

ACES house (9)

The Annual Cycle Energy Systems (ACES) demonstration house is a
2000-ft^2 single-family residence built to demonstrate the energy-
conserving features of added insulation, a ventilation cooling cycle,
and the ACES. Its primary features is the use of the ACES--an integrated heating, cooling, and domestic hot water system that employs
a high-efficiency unidirectional heat pump, low-temperature thermal
storage, and solar assistance.

The ACES house is part of a three-house complex located on the
University of Tennessee Agriculture campus near Knoxville, Tennessee
(see Fig. 7-21). The ACES house, the solar house, and the control
house are in the same complex and have the same floor plan, so that
the three houses exhibit nearly equivalent heating and cooling loads.
The solar house employs a solar heating and hot water system that
supplies some portion of the space and water heating loads. The
control house uses either resistance heat or an air-to-air heat pump
to supply the heating load, an air-to-air heat pump to supply the
cooling loads, and an electric water heater to supply the domestic
hot water loads.

The ACES concept relies on the fact that a heat pump must extract heat from some external source to provide heating. In a conventional heat pump system, this external source is the outside air
accessed by an outdoor air unit. In the ACES, the outside air unit
is replaced by ice-freezing coils located in an ice-and-water

Table 7-15. Construction Cost Estimate[1]

	Materials	Labor	Total
Cost Increases			
Thermal Storage Bed			
Excavate Extra 3 Ft. Deep and Level Stone		$ 190.00	
57.75 Tons Crushed Stone	$ 233.00		
4-In. Drainage Pipe through Stone	248.00	106.00	
Load Sand in Concrete Blocks	12.62	36.00	
88 Pcs. 2-In. ×4 Ft. ×8 Ft. Thermax	1,774.00	193.00	
10 Rolls Duct Tape—Alum. Foil	67.80		
Chimney Base—Air Distribution Box	110.00	288.00	
Sand, 50.89 Yards	72.00		
Aluminum Air Duct System	737.00	224.00	
Second Set Batten Boards	38.38	36.00	
1600-W Electric Resistant Wiring (Option)	88.00		
Sand Resistance Wire Placement		116.00	
Duct Work in House and Fans	632.62	288.00	
	$4,004.42	$1,477.00	$5,481.42
Wall and Roof Insulation			
Taping and Caulking	$ 297.70	$ 128.00	
4-In. Thermax on First Level	1,290.94	160.00	
4-In. Thermax on Rest Walls/Spacers	1,713.60	360.00	
4-In. Thermax on Roof/Spacers	2,067.84	280.00	
	$5,370.08	$ 928.00	$6,298.08
Solar Hot Water Heating			
Daystar Panels	$1,490.00	$ 600.00	$2,090.00
Total Increases			$13,869.50
Cost Decreases			
Eliminate 160,00 Btu Oil Furnace			$2,852.00
Conventional Fiber Glass Insulation			1,761.32
Total Decreases			$4,613.32
Net Cost Increase			$9,256.18[2]
			or $4.02/Sq. Ft.

[1]Cost differences as compared with conventional post and beam house for Edinburg, N. Y., summer 1978.

[2]Brownell did not include the cost of a wood stove or other auxiliary heating in this calculation. A wood stove would add roughly $500 to the net cost increase. This omission is offset in part, however, by inclusion of $204 for the optional 1600-W electric resistance wiring in the sand bed.

Source: Ref. 6.

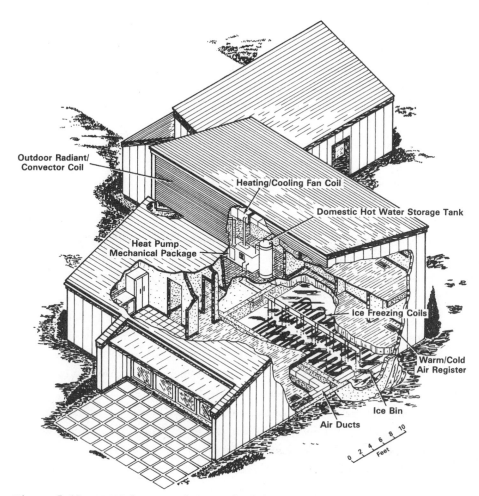

Figure 7-21. ACES house. (From Ref. 9.)

thermal storage bin, and the external source is the heat from the
fusion of water as it is converted to ice.

During the heating season, as the compressor operates to supply
the heating demands, water is gradually converted to ice in the
thermal storage bin. When space cooling must finally be supplied,
the ACES need only melt stored ice to meet the cooling demands. Because the compressor does not operate to meet the cooling loads,

energy savings are obtained by producing the domestic hot water with
the compressor at a coefficient of performance (COP) of approximately
3, as opposed to using electric resistance water heating at a COP
of 1. Being a fixed-capacity system, the ACES further reduces energy consumption by not requiring the use of supplemental resistance
heaters. Consequently, the ACES can deliver all of the space heating at a nearly constant COP of 2.7, even during the coldest days
of winter, whereas a conventional heat pump would drop to a COP of
approximately 1.3 under those conditions.

The ACES, shown in Fig. 7-22, is composed of a compressor,
four refrigerant-to-brine heat exchangers, a fan coil, an outdoor
panel, a domestic hot water storage tank, a thermal storage bin,
and the appropriate piping and valving to interconnect the various
components. The brine in the ACES is a solution of 25% methanol by
volume in water. By employing various valving arrangements, the
ACES has nine distinct modes of operation. Three of these modes
perform the primary functions of space heating, space cooling, and
domestic hot water production; two of the modes perform the balancing functions of ice melting and night heat rejection; and the remaining four modes result from combinations of the previously mentioned modes occurring simultaneously.

When space heating is called for, all pumps, the compressor,
and the fan are in operation, and the only idle component is the
domestic hot water condenser. Space heating is delivered to the fan
coil from the space heating condenser, and domestic hot water is
produced by the superheat obtained from the domestic hot water
desuperheater. The evaporator extracts heat from the thermal storage bin, simultaneously converting some of the water into ice for
future space cooling demands.

The second major function of the ACES, providing domestic hot
water, is automatically performed whenever space heating is called
for, but quite often this amount of hot water is not sufficient.
For those occasions, the ACES is capable of providing domestic hot
water alone. In this mode, the compressor and the hot water and

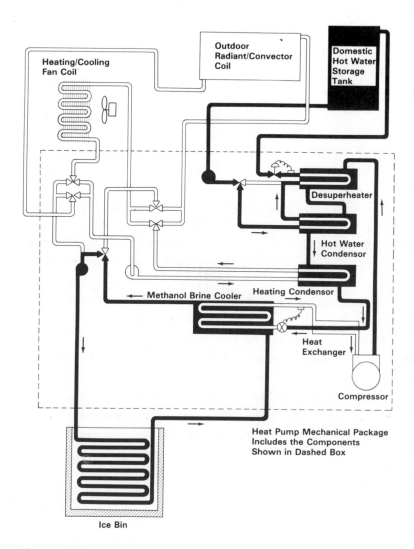

Figure 7-22. Space heating schematic. (From Ref. 9.)

cooling pumps are in operation. Heat, which the evaporator extracts
from the thermal storage bin, forming ice in the bin, is delivered
to the hot water through the domestic hot water condenser and desuperheater acting in series. The ice produced in this mode is also
saved to meet later space cooling loads.

The simplest of the primary modes of operation is that of space
cooling. The fan and the cooling pump are the only electrical components in operation during this mode. Circulating chilled brine
from the thermal storage bin to the fan coil causes some of the ice
in the thermal storage bin to melt, thereby delivering space cooling
to the home.

If a climate were so uniform that the ice produced while delivering space heating and domestic hot water were exactly the amount
required for space cooling, no additional modes of operation would
be needed. However, because this is seldom the case, the ACES must
have some method of balancing its operation. Northern climates necessitate melting some of the ice produced as the ACES supplies the
summer cooling requirement. Southern climates necessitate making
extra ice when the supply has been exhausted. These two external
balancing modes are provided by ice melting and night heat rejection.
Ice melting is accomplished by circulating chilled brine from the
thermal storage bin to the outdoor panel, where heat from the sun
and/or the ambient air, if the temperature exceeds $32^{\circ}F$, raises the
brine temperature and melts some of the ice. The only electrical
device operating during this mode is the cooling pump. Night heat
rejection is accomplished by extracting heat from the thermal storage bin, thereby producing ice, and placing some of this heat into
the domestic hot water; the excess heat is rejected through the
outdoor panel. This mode requires that the compressor and all pumps
be in operation.

The ACES offers two major features that make it more attractive
than conventional heating and cooling systems. For the consumer,
the ACES offers a reduction in the amount of electricity consumed in
providing space heating, space cooling, and domestic hot water.

For the utility, the ACES automatically provides load leveling and peak demand reduction capabilities. This is done remotely, without the utility having to control the heating and cooling systems or having to rely on the consumer to perform voluntary load leveling. With the introduction of time-of-day metering and lower off-peak rates, these load-leveling capabilities will be economically attractive to the consumer as well.

The ACES reduces electrical consumption in a number of ways, one of which is a deferred saving which results from the use of stored ice in providing almost free space cooling. During the period before the ice supply has been exhausted, which varies from climate to climate, the ACES supplies space cooling at a COP of approximately 13, as compared to a COP of approximately 2 to 3 for conventional cooling systems. The result is an electrical saving of 77 to 85% over conventional systems. When the stored ice has been exhausted and night heat rejection is required, the ACES COP for cooling is nearly the same as that for a conventional system. However, the ACES produces domestic hot water under these conditions as a by-product of compressor operation.

Electrical consumption is also reduced during the heating season. This saving is a consequence of the ACES being a fixed-capacity system: The ACES has no need for supplemental electric resistance heating to meet the heating loads on extremely cold days and can maintain a high average COP during the heating season.

Figure 7-23 shows the hourly consumption of electricity for both the ACES and the control house for the week beginning Jan. 2, 1978. Because of the unusually cold weather during this week, the electric consumption of a conventional heat pump system would have been almost equivalent to that of an all-electric system. The plot shows that the ACES reduced the peak power demand from 13 to 3.1 kWh, resulting in a 76% peak power reduction at 8:00 a.m., the most critical time for the utility.

Preliminary estimates indicate that in the Knoxville area the ACES consumes approximately 51% less energy annually than does a

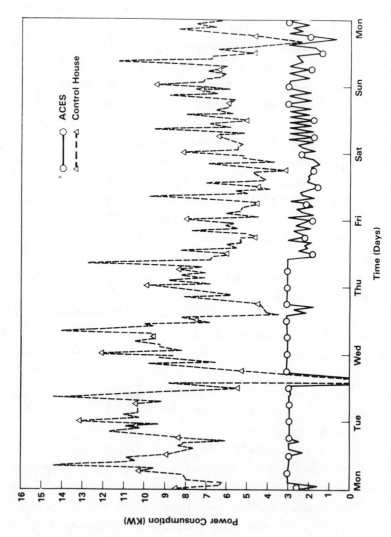

Figure 7-23. Comparison of ACES and control house week beginning 1/9/78.
(From Ref. 9.)

conventional heat pump system with a seasonal COP of 2 and an electric resistance hot water heater, and 63% less energy than does an
electric furnace, a central air conditioning system with a cooling
COP of 2, and an electric resistance hot water heater.

CONCLUSION

Energy-efficient home designs are practical today. However, home
designers must start putting these theories into practice, and home
owners and builders must begin demanding them. As shown in the case
studies, energy efficiency improvements of 50% or more are both
feasible and cost-effective. Every energy-conscious home should approach the conservation of energy through passive solar designs and
must carefully consider building location, configuration, orientation to the sun and wind, layout, method of construction, and other
important design details.

Homes in the past have been designed and built with greater
energy efficiency than those that are being built today: The older
houses took advantage of the natural environment. This is a lesson
we must learn if we are to alleviate our energy woes.

NOTES AND REFERENCES

1. Eugene Eccli, ed., *Low Cost, Energy Efficient Shelter for the Owner and Builder,* Rodale Press, Emmaus, Pa., 1976, pp. 166-170.

2. National Research Council, *Solar Radiation Considerations in Building Planning and Design,* National Academy of Sciences, 1976, p. 108.

3. National Solar Heating and Cooling Information Center, *Passive Design Ideas for the Energy Conscious Architect,* U.S. Department of Housing and Urban Development, August 1978, pp. 37-40.

4. Taghi Alereza and James Barber, A geographical analysis of po- tential energy savings in residential buildings, in *Technology and Energy Conservation,* Information Transfer, Inc., Rockville, Md., June 8-10, 1977, pp. 143-150.

5. Burt Hill and Associates, *Minimum Energy Dwelling Workbook,* U.S. Department of Energy, December 1977.

6. Ralph F. Jones, R. F. Krajewski, and Gerald Dennehy, *Case Study of the Brownell Low Energy Requirement House,* Brookhaven National Laboratory, Upton, N.Y., May 1979.

7. Owens-Corning Fiberglas Corporation, *The Arkansas Story,* Report No. 1, 1978.

8. R. P. Stromberg and S. O. Woodall, *Passive Solar Buildings: A Compilation of Data and Results,* SAND 77-1204, Sandia Laboratories, 1977.

9. A. S. Holman and V. R. Brantley, *ACES Demonstration: Construction, Startup and Performance Report,* ORNL/CON 26, Oak Ridge National Laboratory, Oak Ridge, Tenn., October 1978.

8

Energy Conservation in Commercial Buildings

Kirtland C. Mead

Resource Planning Associates, Inc.
Paris, France, and Cambridge, Massachusetts

The commercial sector includes firms involved in wholesale and retail trade; in finance, insurance, and real estate; in lodging services; in public administration; and in the administration of
transportation, communications, electricity, gas, and sanitary services. Past patterns of energy usage in the commerical sector suggest that it is a fertile area for future conservation efforts.

Commercial energy use for selected years from 1950 to 1975 is
summarized in Table 8-1. As is evident, commercial energy use grew
very rapidly over the pre-Arab embargo period (4.7% from 1950 to
1972), with this rapid growth being even more pronounced for the
period immediately preceding the Arab oil embargo. For example,
energy use in the commercial sector grew at a faster rate than energy
use for the nation, increasing at an annual rate of 6.2% from 1960
to 1972.

The rapid growth and increasing importance of commercial energy
use are the results of a demand for services that increased at a
faster rate than the total output of the economy. A number of changing societal factors contributed to this increasing relative demand.
One such factor is the increase in the number of secondary workers

Table 8-1. Commercial Energy Use, 1950-1975 (QBtu)

Year	Electricity[a]	Gas	Oil	Other[b]	Total	Total as Percentage of National Energy Use
1950	0.90	0.40	0.93	1.14	**3.37**	9.9
1955	1.32	0.61	1.22	0.71	**3.86**	9.7
1960	1.43	0.99	1.64	0.47	**4.53**	10.2
1965	2.43	1.22	1.96	0.43	**6.04**	11.3
1970	3.83	1.77	2.25	0.43	**8.28**	12.3
1971	4.11	1.98	2.12	0.43	**8.64**	12.6
1972	4.46	2.05	2.32	0.45	**9.28**	12.9
1973	4.71	2.03	2.30	0.41	**9.45**	12.6
1974	4.73	2.01	2.06	0.39	**9.19**	12.6
1975	5.04	2.05	1.92	0.37	**9.38**	13.3

[a]Electricity is reported in this study as primary energy, that is, losses in generation, transmission, and distribution are included.

[b]Coal and liquid natural gases.

Source: Jerry R. Jackson, An Economic-Engineering Analysis of Federal Energy Conservation Programs in the Commercial Sector, ORNL/CON-30, Oak Ridge National Laboratory, Oak Ridge, Tenn., January 1979, p. 6.

(working wives) in the labor force, which created an increase in demand for services formerly provided in the home. The recent growth of fast-food establishments is probably the most outstanding example of this phenomenon. Increasing real family incomes also contributed to the increasing economic importance of the commercial sector.

Evidence from numerous conservation case studies indicates a tremendous conservation potential in the commercial sector: Savings of up to 60% have been reported (1). This large conservation potential, along with the rapid growth relative to other sectors, suggests that this sector should receive more consideration than might be accorded it strictly on the basis of the amount of energy used. This chapter describes, in some detail, conservation actions which can be taken to reduce energy consumption in commercial buildings.

THE COMMERCIAL SECTOR

The commercial sector is a catch-all statistical category which includes all nonresidential buildings. Table 8-2 describes the commercial building inventory in the United States. These figures (still the latest reliable figures available) imply a 50% increase in the building stock over the decade 1970-1980, or a compound annual growth rate of 3.5%. This growth rate, which is substantially greater than the population growth, reflects the shift of the U.S. economy into the service sector.

The regional distribution of the commercial building stock follows that of the population (see Table 8-3). Broadly speaking, the greater part of the existing building stock is in the Northeast and North Central regions, areas with cold climates. But much of the newer stock, buildings built in the wasteful 1960s, is probably in the South and Southeast, areas requiring more air conditioning. The 1980 census figures will show a further shift of the stock into the South and West, and we may expect that a more than proportional share of new construction is taking place in these regions.

The data also show that about 60% of the commercial stock consists of offices, retail establishments, and schools--buildings whose primary function is to provide conditioned space for light human

Table 8-2. Commercial Building Inventory (Millions of Square Feet)

	1970	% of Nations Total	1980 Ext.	% of Nations Total
Offices	3,380	16	5,681	17
Retail	4,210	19	7,575	23
Schools	5,040	23	6,804	21
Hospitals	1,500	7	2,218	7
Other	7,480	35	10,458	32
Total	21,650	100	32,645	100

Source: Arthur D. Little, Inc., Residential and Commercial Energy Consumption, Cambridge, Mass., August 7, 1974.

Table 8-3. Commercial Inventory by Region (Millions of Square Feet)

Region and Segment	1970 Inventory	Region and Segment	1970 Inventory
Northeast		**West**	
Offices	913	Offices	676
Retail	1,094	Retail	758
Schools	1,159	Schools	857
Hospitals	435	Hospitals	210
Other	1,800	Other	1,415
Total region	**5,401**	**Total region**	**3,916**
North Central			
Offices	845		
Retail	1,221		
Schools	1,411		
Hospitals	420		
Other	2,093		
Total region	**5,990**		
South		**U.S. Total**	
Offices	946	Offices	3,380
Retail	1,137	Retail	4,210
Schools	1,613	Schools	5,040
Hospitals	435	Hospitals	1,500
Other	2,172	Other	7,480
Total region	**6,303**	**Total United States**	**21,610**

Source: Arthur D. Little, Inc., *Residential and Commercial Energy Consumption*, Cambridge, Mass., Aug. 7, 1974.

activity. Thus, we may expect that the energy systems in these three classes of buildings will be similar, especially in the larger buildings.

In 1975, the commercial sector represented 13.3% of the total energy consumption in the United States, up from 9.9% in 1966. A recent study by Jackson and Johnson, which developed detailed data on energy consumption in the commercial sector, lists their estimates for 1975 consumption by type of commercial building, as shown in Table 8-4 (2). A comparison of these figures with the commercial inventory estimates in Table 8-2 shows that offices and hospitals use more than their fair share of energy, while schools and "other" use less. Hospitals are especially energy-intensive. The retail

Table 8-4.   Commercial Energy Use by Building Type,
1975

Building Type	Energy Use (10^{15} Btu)	Percent of Total Energy Use
Office		
Public administration	.40	4.3
Finance & other office	1.44	15.7
Subtotal	**1.84**	**20.0**
Retail—wholesale	2.20	23.9
Schools & educational	1.77	19.3
Hospitals	1.08	11.7
Other:		
Hotel—motel	.56	6.1
Warehouses	.32	3.5
Religious	.26	2.8
Garages & service sta.	.09	1.0
Miscellaneous	1.08	11.7
Subtotal	**2.31**	**25.1**
Total	**9.20**	**100.0**

Source: Ref. 2, p. 9.

category also uses a slightly larger percentage of the total commercial energy consumption than its relative share of the commercial
inventory would indicate.

Commercial energy use according to type of fuel/end use, is
provided in Table 8-5. If generation and transmission losses were
included, electricity would represent 55% of the total consumption
and natural gas another 22%. The largest fuel/end use components
are oil and gas for space heating (38%) and electricity for cooling
(20%).

Table 8-6 shows how commercial energy consumption by end use
has been changing over time. Since 1965, the relative shares of
space heating and water heating have shrunk, in favor of lighting
and air conditioning. To some extent, this reflects the shift of
the population and the commercial building stock towards cooling-
intensive climates, but there has also been a general raising of

Table 8-5. Commercial Energy Use by Fuel/End Use, 1975 (10^{15} Btu)

	Space Heating	Cooling	Water Heating	Lighting	Other[a]	Total
Electricity[c]	0.33	1.83	0.04	2.09	0.76	5.05
Gas	1.66	0.14	0.08		0.17	2.05
Oil	1.88		0.10			1.98
Other[b]	0.12					0.12
Total	3.99	1.97	0.22	2.09	0.93	9.20

[a]Other end uses include cooking and electromechanical uses.
[b]Other fuels include coal and liquid natural gases.
[c]Includes losses from generation and transmission.
Source: Ref. 2, p. 9.

Table 8-6. Commercial Energy Use by End Use (10^{15} Btu)

Year	Space Heating	Cooling	Water Heating	Lighting	Other[a]	Total
1965	3.124	1.037	0.1797	1.019	0.358	5.718
1966	3.296	1.145	0.1895	1.151	0.426	6.208
1967	3.519	1.261	0.2024	1.234	0.416	6.633
1968	3.650	1.365	0.2097	1.358	0.481	7.064
1969	3.760	1.466	0.2160	1.481	0.543	7.466
1970	3.979	1.605	0.2284	1.608	0.574	7.995
1971	4.149	1.735	0.2376	1.728	0.615	8.465
1972	4.333	1.876	0.2485	1.875	0.660	8.993
1973	4.319	1.945	0.2480	1.979	0.720	9.210
1974	4.050	1.899	0.2321	1.963	0.808	8.952
1975	3,995	1.962	0.2287	2.093	0.927	9.205

[a]Other end uses include cooking and electromechanical uses.
Source: Ref. 2, p. 12.

lighting and air conditioning standards, especially in the years prior to the oil embargo. "Other" consumption has also shown impressive growth.

The importance of the commercial sector for energy conservation in the United States derives as much from its increaisng energy consumption as from the absolute levels of its energy consumption. Driven by long-term shifts in the economy, the commercial sector is increasing its energy consumption share, relative to the residential, transportation, and industrial sectors.

Since it is impossible, in a short chapter, to discuss all the conservation measures that can save energy in all of the different types of commercial buildings and since about 60% of the commercial square footage is in "office-like" buildings (offices, retail establishments, and schools), office buildings are the focus of the rest of this chapter. Our analysis is also applicable to hospitals, which are the most energy-intensive of buildings; however, the host of conservation measures which relate specifically to hospitals are not discussed. In the sections that follow, the typical office building, including its structure and systems, is described. Each component of the building is analyzed, and conservation measures to reduce energy consumption are identified.

THE OFFICE BUILDING AND ITS ENERGY SYSTEMS

Most of the existing commercial building stock was built in the pre-embargo era when energy was cheaper than investment capital. The architect's first aim was usually to minimize cost-per-square-foot to the contractor or owner, so that a minimum was spent on the structure of the building. These trends were exacerbated by the "international style" of architecture, which placed a premium on large windows and clean exteriors.

The result was the typical office building that has extremely poor thermal performance and is extremely sensitive to weather, sun, and wind. In winter, it is common for office buildings to require heating on the north side and air conditioning on the south. Interior areas away from windows may require cooling year round, while perimeter areas will require cooling and heating, depending on outside conditions. Designers relied and continue to rely on complex heating, ventilating, and air conditioning (HVAC) systems to combat these extremes and to maintain tolerable thermal conditions within these buildings.

Commercial buildings in most sections of the country are characterized by their large cooling demands. These demands result from weather but also from the large amount of heat produced by lighting, electrical equipment (e.g., computers), and the inhabitants themselves. These internal loads are often larger than those produced by the weather. Even in the cooler areas of the country, many commercial buildings do not require a central heating plant. Thus, the size of a building's HVAC energy system is usually determined by the demands for cooling on the hottest, most humid summer day.

One of the major loads on any building is caused by infiltration of outside air through cracks around doors, windows, and other elements of the exterior. Infiltration, which in a particularly bad building can account for up to 25% of the entire energy consumption, is one of the most intractable problems in energy conservation in commercial buildings. Infiltration alone can result in one-half an air change per hour in the entire building. It is made worse by the so-called "chimney effect," a term describing the general tendency for air to move vertically in tall buildings, with hot air rising and cool air falling. The chimney effect is promoted by open vertical spaces, such as staircases and elevator shafts, and can only be controlled by sealing openings through which air can enter of leave the building.

From an energy point of view, the HVAC system is the core of a commercial building, and understanding the components of the HVAC system is basic to discussions of conservation in the commercial sector. Because cooling is so often the dominant load, because cooling and heating must often be provided simultaneously, and because of the large areas involved, almost all office buildings are conditioned with both hot and cold air at the same time. Fans deliver air to the interior space at volumes and temperatures required to offset internal loads and thermal losses (or gains) to the exterior and maintain the space at the desired temperature and humidity conditions. The supply air, after space conditioning, is recovered through return air registers and recirculated through the supply fans. A portion of it is exhausted to the outside, and fresh air is introduced to replace the exhaust air. Fresh air intake is usually at a higher rate than exhaust air, to ensure that the building remains at a slight positive pressure. This damps out the variations in load which result from infiltration changes from one side of the building to the other.

The air is conditioned to the proper temperature through the use of heating and cooling coils that are placed in the air ducts between the supply fans and the conditioned space. Filled with hot or cold water supplied by the building's central plant, these coils add or extract heat from the supply air as required, to maintain the space at the desired temperature. The heating and cooling coils are the interface between the air side of the HVAC system and the water side.

HVAC systems differ according to the precise location of the heating and cooling coils in the air stream and according to the manner in which the system adjusts supply air temperatures to meet varying space loads in each of the buildings' zones. Very often, different system types or methods of control will be used in zones subject to different types of loads. It is very common, for instance, to have one type of HVAC system for the interior area (or zone) of the building and another for the perimeter area. The interior zone is far from the windows, is lighted all the time, and often contains the building's service core (e.g., elevators, lavatories and/or the computer); it may have a cooling load 365 days a year. By contrast, the perimeter is influenced by the windows and can have a heating or cooling load, depending on the weather, building occupancy, and the time of day. Thus, different HVAC systems

are often appropriate for these different load patterns. Energy
usage in a commercial building depends critically on how the HVAC
systems in the different zones work together to condition the space,
as heating and cooling loads change over the day.

It is therefore useful to outline the basic types of HVAC systems before discussing how those systems' performance might be improved. The simplest and most common system is the single duct system, which supplies air at a fixed rate to the space (see Fig. 8-1).
As space conditioning loads drop below the design load, output is
varied by changing the temperature of the supply air or by cycling
the system on and off. Most often, heating air is provided at 100
to $120^{\circ}F$, and cooling air at about $55^{\circ}F$.

Fine-tuning space conditions to occupants' needs in different
areas or zones of the building is difficult to do in the single
duct system. Consequently, the single zone constant air volume system with terminal reheat has been developed to provide more flexible
control of temperature and relative humidity (see Fig. 8-2). All
the supply air is cooled to the temperature that is required by the
zone with the largest heating demand; often that temperature is $55^{\circ}F$.
In all other zones, the air is reheated by terminal reheat coils before it is introduced into the space, so that the amount of reheat
can be controlled in each zone individually. While it provides
exemplary climate conditions, this type of system is very wasteful
of energy because it heats and cools the air going to most zones
and can consume up to twice the energy of an efficient HVAC system.
The elimination of simultaneous heating and cooling of supply air
for a single conditioned space should be a major objective of efforts to improve the efficiency of HVAC systems.

Multizone and dual duct systems (Figs. 8-3 and 8-4) are other
HVAC designs that mix hot and cold air to produce supply air at the
right temperature for each of several zones. In these systems, the
opportunities for energy waste are enormous, since the heating and
cooling coils must be maintained at the extreme temperatures required to supply the extremes of heating and cooling demand at the

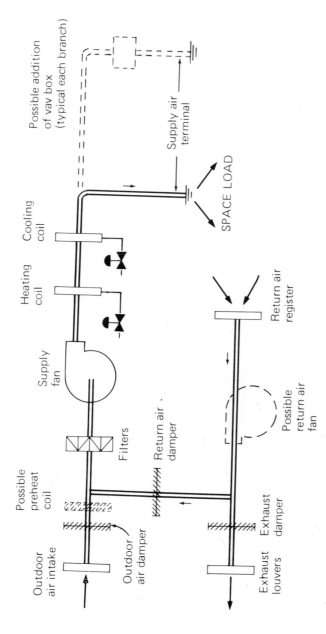

Figure 8-1. Single duct system. (From Fred Dubin, Harold Mindell, and Delwyn Bloome, Guidelines for Saving Energy in Existing Buildings: Engineers, Architects, and Operators Manual, ECM 2, Federal Energy Administration, June 16, 1975, p. 215.)

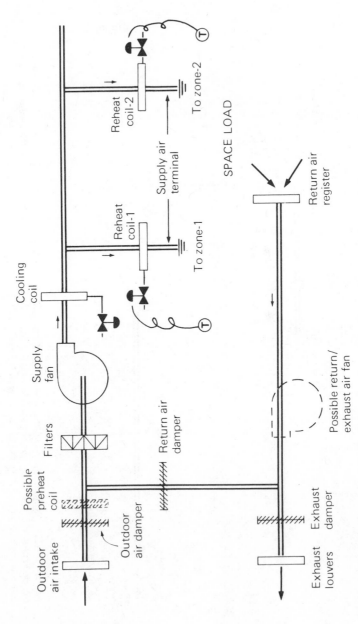

Figure 8-2. Terminal reheat system. (From F. Dubin, H. Mindell, and D. Bloome, Guide-lines for Saving Energy in Existing Buildings: Engineers, Architects, and Operators Manual, p. 216.)

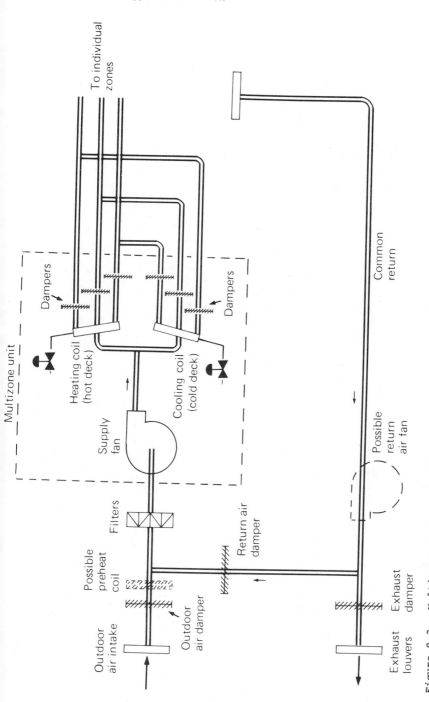

Figure 8-3. Multizone system. (From F. Dubin, H. Mindell, and D. Bloome, Guidelines for Saving Energy in Existing Buildings: Engineers, Architects, and Operators Manual, p. 218.)

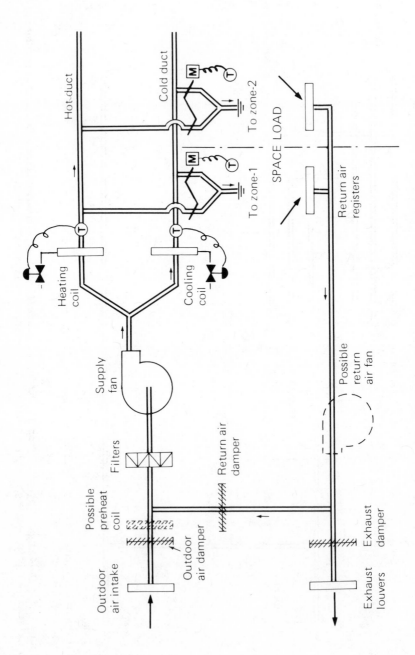

Figure 8-4. Dual duct, low-velocity system. (From F. Dubin, H. Mindell, and D. Bloome, Guidelines for Saving Energy in Existing Buildings: Engineers, Architects, and Operators Manual, p. 220.)

same time. Energy is wasted whenever conditioned air is supplied to zones whose demands lie between these extremes.

The variable air volume (VAV) system (Fig. 8-5), is another adaptation of the single zone, single duct system. In the VAV system, the demands of different zones are met by varying the amount of air, which is at a constant temperature, to the respective zones. Air volume can be varied either by leaving the supply fan volume constant and bleeding off unused air, through dump boxes, directly into the return air registers, or by designing the dump boxes to regulate the entire airflow to the space. In either case, fan volume must vary with space load. Used primarily in zones which require year-round cooling loads, VAV systems are much more energy-efficient than terminal reheat, multizone, or dual duct systems. One major HVAC conservation measure entails the introduction of VAV load control into existing HVAC systems that do not already have it.

A final design that is often used to condition perimeter zones with large fluctuations of temperature is the induction unit system (Fig. 8-6). Primary air from the supply fan is either heated or cooled and is supplied at high velocity and pressure to induction units in the space. The high velocities induce air in the space to circulate through the induction unit, where it is conditioned by individual water coils. Once again, this is a system that can waste energy by cooling and then reheating the same air.

The previous discussion has been on the air side of the different HVAC systems; however, variation is also possible on the water side. Fan coil units in all of the designs can be supplied by two-pipe, three-pipe, or four-pipe water distribution systems. The two-pipe system, in which either hot or cold water is supplied to a single coil, can only heat or only cool at any given time. Manual changeover is required to switch from heating to cooling, so that system performance can be poor in fall and spring, when the building could require heating and cooling at the same time. Most modern systems are therefore three- or four-pipe systems, which permit simultaneous heating and cooling of different zones. In a three-pipe system, however, the return pipe is common to both the

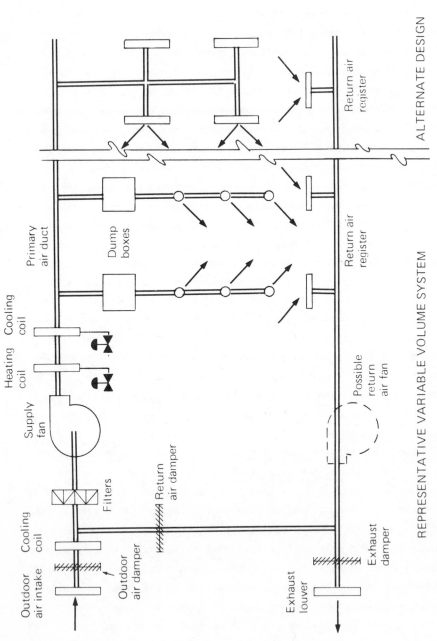

REPRESENTATIVE VARIABLE VOLUME SYSTEM

Figure 8-5. Variable volume system. (From F. Dubin, H. Mindell, and D. Bloome, Guidelines for Saving Energy In Existing Buildings: Engineers, Architects, and Operators Manual, p. 222.)

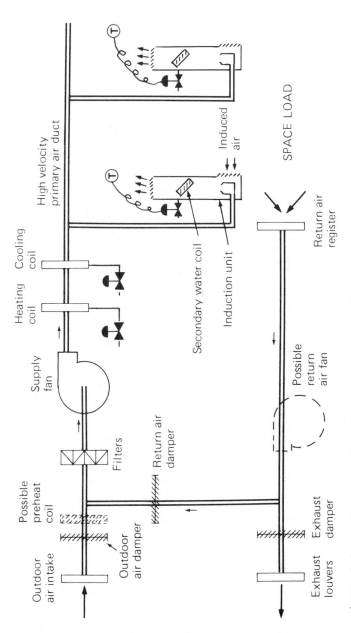

Figure 8-6. Induction unit system. (From F. Dubin, H. Mindell, and D. Bloome, Guidelines for Saving Energy in Existing Buildings: Engineers, Architects, and Operators Manual, p. 224.)

heating and cooling coils, and this results in the same sort of wasteful mixing of hot and cold fluids that was previously discussed.

HVAC designs can exist in a wide variety of sizes. In smaller sizes, up to 20 tons, the central chiller plant can be located in or near the conditioned space and can condition air directly. Since no water distribution system is needed, this is called a direct expansion system. Slightly larger "rooftop" units exist in capacities of up to 60 tons. Comprised of a furnace, refrigeration compressor, condenser, and fan-coil unit, these modular or packaged systems have begun to compete with custom-designed central plants for the larger building market. Such units offer economies in distribution, because less water piping and air ducting is required and because much less fan power is required. However, the modular plant itself is not nearly as efficient as a central plant. According to Fred Dubin, a leading expert on HVAC, a central HVAC plant is 10 to 15% more efficient than a packaged plant. A major reason for this is that the condenser of the central plant chiller releases heat to water (usually a cooling tower), while the packaged plant condenser usually transfers heat to the outside air, in the manner of a window air conditioner. However, water is a much more efficient heat transfer medium than air, and much more fan power is required to operate an air-cooled condenser than to operate a cooling tower. There are also new high-pressure steam absorption chillers that can compete, on an energy-efficiency basis, with electric equipment.

The next section discusses various conservation measures in the average office building. Each component of the office building is considered, and ways to save energy in it are explained. Logically, we must begin with ways to reduce space conditioning loads and then discuss how best to build or modify the HVAC system to meet these reduced loads. We begin with a discussion of the building shell, through which loads are transmitted as a result of temperature differences, wind (primarily infiltration), and sun. We then consider other sources of space conditioning loads, such as lighting and ventilation with outside air. Finally, we consider the HVAC system itself, ending with a discussion of the central plant and of the controls for the HVAC system as a whole. The measures discussed are not exhaustive; they are merely some of the most important ways to save energy. Many other areas for conservation, including those of sanitary hot water and electric motors and equipment, are not discussed.

All the conservation measures discussed would "add" to a total reduction of energy consumption in the building. Improvements in HVAC alone can deliver savings on the order of 15%, and with all measures taken together, 30% or more. But it is impossible to describe in any simple way how the various energy savings measures might add together in any particular building, let alone in the typical office building treated. Each HVAC system, like every commercial building, is a custom design located in a unique microclimate. Energy-efficient design or retrofit cannot be accomplished by using cookbook recipes or simple rules. A detailed simulation of the system's performance is required in order to assess the potential effect of any change in design. Sophisticated computer programs exist for this purpose, and these are discussed in the last section.

THERMAL PERFORMANCE OF THE BUILDING SHELL

Because most existing office buildings were designed for minimum first cost, the result has been a light and thermally inefficient building shell. These buildings' walls, roofs, and windows can often be made substantially more energy-efficient at low capital cost to the owner. In addition, careful siting and orientation of the building can result in large savings at zero or very low cost. In this section we review such measures in detail and provide recommendations for improving energy efficiency.

Siting

Orientation of a building can influence energy consumption significantly. If the building is rectangular, it should be sited with the
long axis east-west, since east and west walls receive more direct
sunlight than the south wall, and for longer periods of time. In
buildings that require cooling, wall and glass exposure should be
minimized towards the west and southwest. The impact of solar heat
gain on the building can be reduced substantially by locating unconditioned spaces on this side. Particularly in colder climates,
northern walls should have few windows and should be tightly sealed
against infiltration losses. With smaller buildings, plantings
should be used to shade the east and west facades in summer and slow
wind speeds near the walls in winter. The latter is especially
important. Plantings near the walls and below the windows can "anchor" a layer of air near the wall and help insulate the building.
The same effect can be obtained with clinging vines.

Building Shape

Much of the commercial construction in the 1950s and 1960s was open-
plan and single-story. The result was a great many buildings with
large surface areas in relation to their internal volume. In general, especially for smaller buildings, compact, multistory shapes
should be chosen, since a three-story building has 35% less surface
area than a single-story building of equivalent floor area. As the
building grows in size, however, there may be an increasing problem
in eliminating the heat generated by internal loads such as inhabitants, lighting, and data processing. Thus, careful analyses must
be performed in order to balance the volume to surface ratio and
the impact of internal loads.

Improved Resistance to Heat Transfer

Within a given building shape, much can be done to limit the heating and cooling loads transmitted to the interior. In a new building, it is generally a good idea to increase the mass of the con-

struction, since greater mass increases the thermal storage capacity of the building and permits the structure to smooth out sudden thermal load changes caused by weather, wind, and sun. Large thermal mass reduces energy consumption of the building, damps out thermal peaks, and reduces the size of the central HVAC plant needed. This is especially true for roofs. A roof of great enough mass (80 to 90 lb/ft^2) can reduce, by 27%, the annual heat loss produced by a lighter roof (of 10 to 12 lb/ft^2). Even more energy can be saved simply by spraying or covering the roof with water in the summer, because a water-covered roof gains 50 to 80% less heat than a dry, dark roof. Similar effects can be obtained by choosing a roof color carefully: it should be light in a hot climate, dark in cold ones.

The thermal quality of the roof should be evaluated, and additional insulation considered. Exterior insulation may be cost-effective if the roof needs repair. In some cases, insulation can be applied to the underside of the roof, but the practicality of this depends on ease of access to the roof and on the presence of other structures, such as ductwork. The plenum area, between the drop ceiling and the roof above, often provides easy access to the underside of the roof, where insulation can be installed. If there are impediments to the fitting of rigid insulation, foam insulation techniques may work.

In slab-on-grade buildings, significant heat loss (or gain) occurs at the perimeters of the floor. Insulation can be added to the outside wall, and it is even beneficial to extend the insulation down the sides of the foundation into the ground. Figure 8-7 depicts insulation of the floor perimeter. Fred Dubin indicates that insulation of this type is economic if condensation occurs on the floor perimeter in cold weather or if floor temperatures close to the outside wall are $10^{o}F$ lower than in the rest of the space. It may also be economic to insulate the underside of suspended floors, if they are in contact with the outside air or with unconditioned space, such as garages or warehouse space.

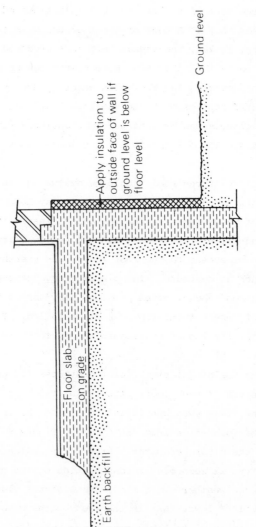

Figure 8-7. Retrofit insulation installation to on-grade floors. (From F. Dubin, H. Mindell, and D. Bloome, Guidelines for Saving Energy in Existing Buildings: Engineers, Architects, and Operators Manual, p. 61.)

The incentives to insulate exterior walls are much the same as those for roofs and floors. In cold climates, particular attention must be paid to north walls. Exterior insulation of walls, while a growing practice in both the United States and Europe, is complicated by aesthetic issues and by window and door openings. If insulation is impossible, it may at least be possible to rusticate exterior walls to increase their resistance to airflow. As in the case of wall plantings, this increases the depth and efficiency of the insulating air film that remains more or less stationary near the wall. Interior insulation is likely to be more complicated than in the case of roofs and ceilings, because there is no plenum space to facilitate access.

Insulation applied to floors, roofs, and walls must have vapor barriers to prevent moisture from condensing in the insulation and ruining its insulating qualities. The optimal amount of insulation depends on the weather and on other efficiency improvements one may be considering for the particular building, and the calculation of this amount usually requires a detailed simulation of the building in its microclimate. These complexities notwithstanding, the Department of Energy has published the recommended thermal performance values presented in Tables 8-7 and 8-8.

Table 8-7. Recommended Insulation Value for Heat Flow Through Opaque Areas of Roofs and Ceilings

Heating Season	U Value (Btu/hr/sq ft/° F)
Degree Days	
0–1000	0.12
1001–2000	0.08
2001 and above	0.05

Source: U.S. Department of Energy, *Instructions for Energy Auditors,* vol. II, EOD/CS 0041/13, Superintendent of Documents, U.S. Government Printing Office, WA, D.C. 20402, September 1978.

*Table 8-8. Recommended Insulation Value for Heat
Flow Through Opaque Exterior Walls for Heated Areas*

Heating Season	U Value (Btu/hr/sq ft/°F)
Degree Days	
0–1000	0.30
1001–2500	0.25
2500–5000	0.20
5000–8000	0.15

Cooling Season—The recommended U value of insulation for heat flow
    through exterior roofs, ceilings, and walls should be
    less than 0.15 Btu/hr/sq ft/°F.

Source: U.S. Department of Energy, *Instructions
for Energy Auditors,* vol. II, EOD/CS 0041/13,
Superintendent of Documents, U.S. Government
Printing Office, WA, D.C. 20402, September 1978.

These recommendations must be interpreted with caution. Additional insulation may reduce the heating load, but it may not reduce the demand for space cooling, especially in locations with
moderate summer conditions. While insulation reduces heat gain during the day, it may inhibit natural cooling during the night, when
the outside temperature drops below that of the building. Heat not
dissipated naturally must be removed by the HVAC system, increasing
energy use. Additional insulation increases the number of heating
hours during which a building can heat itself with internal loads
(e.g., people, lighting, etc.), but additional insulation can also
raise the number of hours that the cooling systems must run during
the year, despite a lower outside temperature. For locations with
extremely hot summers, these problems do not arise, since the outside temperature is usually above that of the building, even at
night. In such climates, additional insulation reduces cooling demand, just as it reduces heating demand in cold climates.

Windows

The traditional taste for large window expanses is one of the chief
reasons for the poor thermal performance of the commercial building,
since windows can account for 20 to 30% of the heat loss and gain of

that building. Solar heat gain is greatest through east, west, and
south windows, and direct and diffuse radiation represents the
majority of the gain; conduction gains are small by comparison (3).
Table 8-9 shows the magnitudes of solar gain that occur through
east, west, and south windows in various parts of the United States.

As with insulation, double- or triple-glazing has advantages
and disadvantages. Lower U-values limit solar gains in summer, but
they also limit solar gains in winter when they are usually desirable. Double-glazing passes about 10% less light than single-glazing, but because in most climates this loss is more than made up by
reductions in conduction heat losses, double-glazing is worthwhile.
Fred Dubin recommends double-glazing if heating degree-days exceed
1500 per year, and triple glazing if degree-days exceed 7500 annually. Double-glazing is almost never worthwhile in cooling-only
situations.

Another efficient measure for windows is the use of reflective
coatings to block incoming solar radiation. Transmission and reflection is controlled by the thickness of the coating, and glass with
transmittance ratings varying from 8 to 26% can be manufactured.
In a recent study, Libbey-Owens-Ford compared the performance of a
building fitted with chromium-coated, dual-wall insulating glass
against that of a building equipped with conventional 1/4-in. plate
glass. The building with the metallic insulating glass exhibited
much lower thermal peak loads, with the result that the cooling plant
could be 64.7% smaller and the heating plant 53.2% smaller. In another case example, an office building in Miami demonstrated 15%
annual energy savings after being retrofitted with reflective window films.

Interior and exterior solar control devices can be very effective in limiting unwanted solar gain through windows. Interior
shading devices, such as draperies and venetian blinds, cost less
than exterior devices, but they are also less effective, since the
sunlight has already entered the building. Even if the radiation
were intercepted, very little of it would be reflected back out the
window. Most of the radiation would simply be transformed into
thermal energy and load within the space.

Table 8-9. Yearly Heat Gain per Square Foot of Single-Glazing and Double-Glazing

City	Latitude	Solar Radiation Langley's	D.B. Degree Hours Above 78°F	Heat Gain Through Window Btu/Ft²/Year					
				North		East and West		South	
				Single	Double	Single	Double	Single	Double
Minneapolis	45°N	325	2,500	36,579	33,089	98,158	88,200	82,597	70,729
Concord, N.H.	43°N	300	1,750	33,481	30,080	91,684	82,263	88,609	76,517
Denver	40°N	425	4,055	44,764	39,762	122,038	108,918	100,594	85,571
Chicago	42°N	350	3,100	35,595	31,303	93,692	83,199	87,017	74,497
St. Louis	39°N	375	6,400	55,242	45,648	130,018	112,368	103,606	85,221
New York	41°N	350	3,000	40,883	35,645	109,750	97,253	118,454	102,435
San Francisco	38°N	410	3,000	29,373	28,375	88,699	81,514	73,087	64,169
Atlanta	34°N	390	9,400	59,559	50,580	147,654	129,391	106,163	87,991
Los Angeles	34°N	470	2,000	47,912	43,264	126,055	112,869	112,234	97,284
Phoenix	33°N	520	24,448	137,771	97,565	242,586	191,040	211,603	131,558
Houston	30°N	430	11,500	88,334	72,474	213,739	184,459	188,718	156,842
Miami	26°N	451	10,771	98,496	71,392	237,763	203,356	215,382	179,376

Source: Fred Dubin, Harold Mindell, and Selwyn Bloome, Guidelines for Saving Energy in Existing Buildings: Engineers and Architects and Operators Manual, ECM 2, Federal Energy Administration, June 16, 1975, p. 129.

In defiance of theinternational style, commercial buildings are beginning to adopt the exterior shading devices that have, for a long time, been used by residential buildings. An "eyelid," which can be placed over a south window to shade it from the strong, high sun of summer, is one such device. In winter, when the sun's angle is lower and warmth is wanted, solar energy can reach the window under the lid. Vertical fins or louvers can also be fitted to the outside of the building to admit sunlight at some angles but not at others. Louvers are particularly effective on east and west windows that must cope with low morning and evening sun angles. Finally, trees can be effective shading devices, especially on east and west walls. If deciduous, they will block the sun in summer but not in winter, when the sun's rays are desirable. Figure 8-8 shows the deployment of some of these devices.

One of the most effective exterior shading devices is the old-fashioned shutter. In winter, shutters can reduce thermal loss at night, when the building is unoccupied; in summer, they can reduce solar gains during the day. A number of companies in southwestern United States are beginning to market automatic shutter devices that open and close in response to light levels or thermal imbalances. But before accepting such complex solutions, American architects would do well to study the Mediterranean shutter designs that are used even today on commercial office buildings in Naples and Rome (see Fig. 8-9). The articulated design blocks direct sunlight but admits diffuse light, and the shutter can be turned to track the sun. Combined with the immense thermal mass of the buildings, these shading devices can produce tolerable interior conditions, even on the hottest days, with no air conditioning at all.

Temperature Levels

Control of interior temperature and humidity is an absolute necessity for reducing energy consumption of commercial buildings. Improved management of existing temperature controls can alone result in a 10 to 15% energy savings (4, pp. iv-22). For example, Fred Dubin

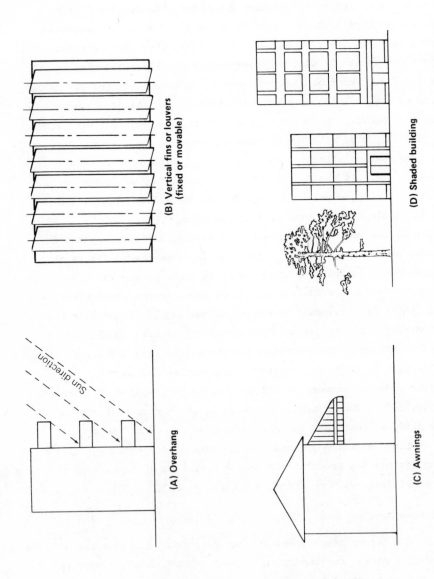

(A) Overhang

(B) Vertical fins or louvers (fixed or movable)

(C) Awnings

(D) Shaded building

Sun direction

Figure 8-8. Solar screening devices. (From Interim Design Criteria: Technical Guidelines for Energy Conservation in New Buildings, U.S. Naval Facilities Engineering Command, January 1975, p. 1-14.)

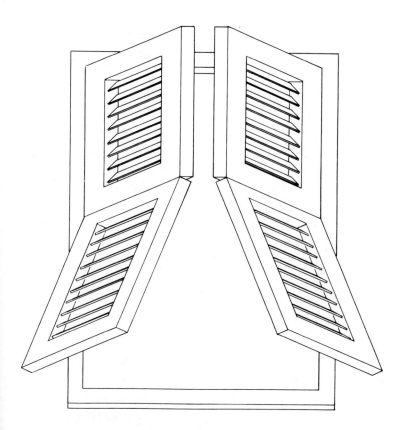

Figure 8-9. Mediterranean shutter design.

predicts that changing interior conditions from $74^{\circ}F$ and 50% relative humidity to $78^{\circ}F$ and 55% relative humidity can save 13% of the energy used for cooling. In the heating season, a temperature of $68^{\circ}F$ should be maintained, but attainable temperature levels will depend on the air tightness of the building, the activity of building occupants, and the relative humidity. Temperatures could be lower if employees would wear light sweaters in the office; provided that they were not subjected to drafts. For those specific locations in the building, such as the areas adjacent to doors and large windows, that are cooler than the average room temperature, it is usually more efficient to provide "spot" heating, rather than

to raise the temperature level of the entire building. The use of
open-plan interiors will reduce the number of cold spots and help
damp out local temperature variations.

An increase in humidity level will usually permit the thermostat to be lowered without decreasing the comfort level of the occupants, but savings are affected by the energy required to humidify
the air. According to Dubin, the energy saved by lowering the
thermostat in a building with small infiltration losses will exceed the costs of humidifying to about 20% relative humidity. Recommended minimum relative humidity levels for winter conditions
are in the range of 20 to 30%.

Every effort should be made to limit the amount of time that
the central space conditioning plant must be operating. If the
building has enough thermal mass, it may be possible to shut down
the heating or air conditioning plant toward the end of the day,
without too drastically affecting the space conditions before quitting time. At night, the building temperature level should be allowed to "float" up or down, as long as it remains within tolerable
environmental limits. Morning start-up (usually a 2-to-3 hr period)
can probably be delayed and spread out, so that the building only
reaches full operating conditions after the start of business.
Since bringing the building to operating conditions after night
shutdown usually produces the largest daily demands on the HVAC
system, retarding morning start-up can save energy and reduce the
size of the required central plant. Nighttime setback and plant on-
and-off control should be automatic and regulated by a 7-day, day-
night clock thermostat. This system could be part of a more comprehensive control system for the building.

Most commercial buildings contain space that is either rarely
inhabited or inhabited only by active people for short periods.
Such spaces should be heated to lower temperatures or not heated
at all. A prime example is stairwells. Since they are used only
by people walking quickly from one heated space to another, they
can be heated to only 55°F. Storerooms and garages may require
little or not heating at all.

LIGHTING

Lighting represents 23% of commercial energy use, and commercial buildings have been notorious for their inefficient and oversized lighting systems. Wasteful lighting designs were stimulated by the desire first to reduce cost and to produce good working conditions. Very often, the classic open-plan office building is lighted by fluorescent lights which cover all or most of the ceiling. Lighting levels are very high over the entire floor area, and there may be only one switch for the entire floor.

Overlamping is often defended on the grounds that lighting load contributes to the heating of the building in winter. This is true, but the same amount of heating can be achieved through the central plant, with much lower capital and operating costs. Overlamping can never be justified primarily as a contribution to heating and, of course, it increases energy use for cooling during the summer.

To reduce energy waste in lighting, lighting levels should first be reduced overall. A recent study by the Environmental Protection Agency predicted that in many buildings, lighting levels could be reduced to 50%, without impairing comfort. Light levels should then be tailored to the task requiring the light ("task lighting"). Some tasks are more difficult than others and thus require more light. The visual difficulty factor (VDF) is defined by the General Services Administration (GSA) as the product of a visual difficulty rating (R) times the number of daily hours that the task must be performed, times a correction factor of 1.5 if the employee is over 50 years of age. Difficulty ratings are listed in Table 8-10. The GSA's recommended lighting levels for task lighting are shown in Table 8-11. The GSA has also recommended general lighting levels for different types of rooms (see Table 8-12). Task lighting may actually increase working efficiency, since it produces a working environment with less glare and more contrast. Energy savings can reach 15 to 20% of the energy presently used for lighting (4, pp. iv-28).

Table 8-10. Visual Difficulty Rating (R) of Tasks

Task Description	Visual Difficulty Rating (R)
Large Black Object on White Background	1
Book or Magazine, Printed Matter, 8-Point and Larger	2
Typed Original	2
Ink Writing (Script)	3
Newspaper Text	4
Shorthand Notes, Ink	4
Handwriting (Script) in No. 2 Pencil	5
Shorthand Notes, No. 3 Pencil	6
"Washed-Out" Copy from Copying Machine	7
Bookkeeping	8
Drafting	8
Telephone Directory	12
Typed Carbon, Fifth Copy	15

Source: General Services Administration, *Energy Conservation Guidelines for Existing Office Buildings,* Superindenent of Documents, U.S. Government Printing Office, WA, DC 20402, February 1977, pp. 4-14.

Table 8-11. GSA-Recommended Lighting Levels

Task or Area	Visual Difficulty (VDF)	Design Level (FC)	Average Level Range (FC)
Service or Public Areas	—	15	12–18
Circulation Areas within Office Space, but not at Work Stations	—	30	24–36
Normal Office Work, Reading, Writing, etc.	1–39	50	40–60
Office Work, Prolonged, Visually Difficult or Critical in Nature	40–59	75	60–90
Office Work, Prolonged, Visually Difficult and Critical in Nature	60 & Up	100	80–120

Source: General Services Administration, *Energy Conservation Guidelines for Existing Office Buildings,* Superintendent of Documents, U.S. Government Printing Office, WA, DC 20402, February 1977, pp. 4-14.

Table 8-12. General Lighting Levels

Areas	Design Level (FC)	Range (FC)
Auditoriums	30	20—40
Cafeteria	30	20—40
Conference Rooms	30	25—35
Corridors, Lobbies and Means of Egress	15	10—18
Kitchen (Average)	50	30—70
Mechanical Rooms (General Areas)	10	5—15
Storage Areas (General Storage)	10	5—15
Storage Areas (Fine Details Required)	30	25—35
Toilets	20	15—30

Source: General Services Administration, *Energy Conservation Guidelines for Existing Office Buildings,* U.S. Government Printing Office, WA, DC, February 1977, pp. 4-14.

At constant illumination (footcandle) levels, considerable energy can also be saved by installing more efficient lamps and lighting reflectors. Figure 8-10 shows that fluorescent and mercury lamps can be twice as efficient as incandescent lamps. The degree to which these lamps can be introduced will depend on the willingness of employees to accept their different light quality.

Lights should be wired so that they can be turned on over only part of a floor. This reduces waste after hours, when commercial buildings are being cleaned, or on weekends, when only a few people are present. Timed switches can be installed to turn lights out automatically in infrequently used areas. This kind of system has been in use for decades in European stairwells.

Finally, maintenance is very important for avoiding waste in lighting. Lighting reflectors and fixtures must be cleaned regularly. Lamp efficiency degenerates over the life of a lamp, and lamps must be replaced before their efficiencies fall too low. All these maintenance measures become much more important if commercial buildings are not overlamped.

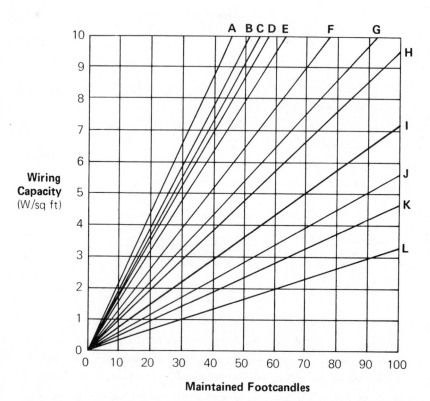

Figure 8-10. *Illumination versus power density. Approximate wiring capacity to provide a given maintained level of illumination in a room of 2.5 room cavity ratio by means of the following: A, indirect, incandescent filament (silvered bowl); B, direct, incandescent filament (with diffuser); C, direct, incandescent filament (downlight); D, general diffuse, incandescent filament; E, direct, incandescent filament (lens); F, direct, incandescent filament (industrial)/indirect, fluorescent (cove); G, indirect, fluorescent (extra high output); H, direct, fluorescent (extra high output, louvered); I, direct, fluorescent (louvered); J, luminous ceiling, fluorescent; K, direct, fluorescent (lens)/direct, HID (mercury); L, direct, semidirect, fluorescent (industrial). (From General Services Administration, Energy Conservation Guidelines for Existing Office Buildings, Superintendent of Documents, U.S. Government Printing Office, WA, DC 20402, February 1977, p. 4-3.*

VENTILATION AND HEAT RECOVERY

Because energy must be expended to condition air brought into the
building from the outside, ventilation represents a significant portion of a commercial building's energy use. John K. Henderson reports that each 1000 cfm of outside air can require 4 tons of air
conditioning capacity (5). Loads introduced by ventilation are similar to those imposed by infiltration, but it much more efficient to
ventilate with the HVAC system than to rely on infiltration.

As is the case with lighting, ventilation standards in the
United States have had little to do with the real needs of the inhabitants. Standard practice has specified ventilation levels much
higher than are really needed for most of the activities in the
building. As with lighting, efficiency can be improved by lowering
the overall level of ventilation and by adapting ventilation levels
to specific tasks.

Table 8-13 provides recommended ventilation levels for different types of spaces. Taken from the American Society of Heating,
Refrigeration, and Air Conditioning Engineers (ASHRAE) conservation
building code 90-75, these levels are still generous, considering
that only 1 to 2 cfm/occupant is required to support physical needs
and that 4 cfm/occupant is ample to control odors and even light
smoking. The noticeable decline in smoking in some offices could,
by itself, be justification for lowering ventilation rates. Local
building codes may also allow many ·spaces to go unventilated, such
as warehouses, stairwells, and entryways. Certainly, if there is
unavoidable infiltration (as in entry ways or around doors), this
may justify reduced ventilation levels.

In general, it is desirable to shut down ventilation during unoccupied hours, and it is usually possible to run the first hour
in the morning and the last hour before quitting time without any
forced ventilation. Indeed, eliminating ventilation during the
first morning hour reduces peak load·on the central plant and may
help reduce the required plant size.

It may, however, be desirable to run the ventilation system at
night, if the days require cooling, but nighttime temperatures fall

Table 8-13.   Ventilation Requirements for Occupants

Application	Ft^3/min Per Person	Application	Ft^3/min Per Person
Banking space	7	Laboratory	15
Barber shop	7	Office	
Beauty parlor	25	General	15
Bowling alley	15	Conference room	25
		Waiting room	10
Cocktail lounge, bar	30	Pool or billiard room	20
Department store			
Retail shop	7	Restaurant	
Storage area serving sales area	5	Dining room	10
Drug store		Kitchen	30
Pharmacists' workroom	20	Cafeteria, short order, drive-in	30
Sales area	7	School[3]	
Factory[2, 3]	10-35	Classroom	10
Garage		Laboratory	10
Parking	1.5[4]	Shop	10
Repair[5]	1.5[4]	Auditorium	5
Hospital[3]		Gymnasium	20
Single or double room	10	Library	7
Ward	10	Office	7
Corridor	20	Lavatory	15
Operating room[6]	20	Locker room[7]	30
Food service center	35	Dining room	10
		Corridor	15
Hotel		Dormitory bedroom	7
Bedroom	7		
Living room (suite)	10	Theater	
Bath	20	Lobby	20
Corridor	5	Auditorium	
Lobby	7	Smoking	10
Conference room (small)	20	No smoking	5
Assembly room (large)	15	Restroom	15
Public restroom	15		

[1] Data extracted from *ASHRAE Standard 62-73*.   Minimum values used.

[2] Special contaminant control systems may be required.

[3] State or local codes are usually determining factor.

[4] Ft^3/min per square foot of floor area.

[5] Where engines are run, positive engine exhaust withdrawal system must be used.

[6] All outside air often required to avoid hazard of anesthetic explosion.

[7] Special exhaust systems required.

Source: Ref. 6.

below those within the building. In such a situation, the ventilation can remove heat load stored in the building during the day and
even "precool" the building in preparation for the following day.
It is more efficient to remove heat this way than to use the central
plant chillers.

The quantity of air exhausted to the outside often exceeds the
minimum outdoor air intake requirement. Additional outside air
must then be introduced to keep the building at positive pressure.
Major economies can thus be made by recirculating a portion of the
exhaust air back into the conditioned space. Figure 8-11 shows how
this can be done. An activated carbon filter is usually introduced
into the recycling duct to remove odors.

A great deal of outside air is required to replace air vented
through kitchen and laboratory hoods. Significant economies are
possible if the replacement air can be introduced directly into the
hood and not into the central HVAC system. The air need only be
conditioned to 55°F in winter and not at all in summer. As shown
in Fig. 8-12, the hoods themselves can be redesigned to operate with
less exhaust, by installing baffles to increase airflow rates in
the hood.

Exhaust air leaves the building at the temperature and humidity
of the conditioned space. In winter, its heat and humidity are pure
energy losses to the building; in summer, the loss is represented
by the energy required to condition outside air to the inside condition. If ventilation volumes exceed 10,000 cfm, and if exhaust
outlets are big enough (e.g., 5000 cfm/location), it would certainly
be worthwhile to install some kind of heat recovery system between
the exhaust and incoming air streams in order to recapture some of
these losses. A number of heat exchangers are available for this
purpose; the designs vary, depending on whether they recapture only
sensible, or sensible and latent heat. If the outside air and exhaust ducts are not colocated, only sensible heat can be exchanged.
A pair of heat exchanger coils may be placed in both the exhaust
and intake ducts and connected to each other by piping filled with

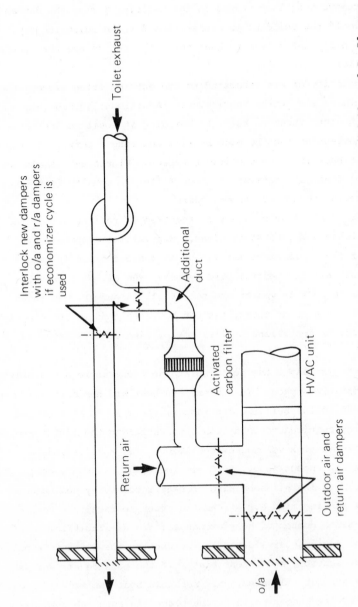

Figure 8-11. Retrofit for recycling exhaust air. (From F. Dubin, H. Mindell, and D. Bloome, Guidelines for Saving Energy in Existing Buildings: Engineers, Architects, and Operators Manual, p. 73.)

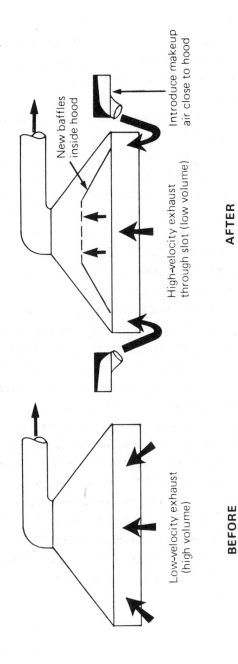

BEFORE

Low-velocity exhaust (high volume)

AFTER

New baffles inside hood

Introduce makeup air close to hood

High-velocity exhaust through slot (low volume)

Figure 8-12. Retrofit of kitchen exhaust hoods. (From F. Dubin, H. Mindell, and D. Bloome, Guidelines for Saving Energy in Existing Buildings: Engineers, Architects, and Operators Manual, p. 75.)

an ethylene glycol solution. The solution is then pumped around the
system to exchange the heat. A heat pipe, a reversible thermal device which consists of a tube filled with vaporizable working fluid,
can function similarly. However, a heat pipe requires colocation
of the exhaust and outside air duct.

The most common device for exchanging both sensible and latent
heat is an "enthalpy wheel" (see Fig. 8-13), in which a mechanical
wheel rotates between the duct containing hot humid air (outside
air in summer, exhaust in winter) and the duct containing cold dry
air (exhaust in summer, outside air in winter). The wheel picks up
heat and moisture from the hot duct and transfers it to the cold.
Up to 90% of the energy normally wasted in the exhaust can be transferred back to the space requiring conditioning.

Reductions in ventilation rates, adaptation of ventilation
rates to different activities, and heat recovery can save a great
deal of energy in a typical office building. In one demonstration,
merely reducing the infiltration rate from 60 cfm/person to 10 cfm/
person resulted in a 17.5% reduction in the annual energy consumption of the building (6). This result is broadly consistent with
estimates made by Mathematica, Inc. in 1975 (4). The major savings
are in heating demand and accrue to fossil fuel.

Before turning to conservation measures for the HVAC system,
we should mention that heat pumps can also recover the waste heat
in the ventilation exhaust. This application of heat pumps is occurring now in Europe, particularly in France. Engineers are installing small heat pumps to recover the heat in exhaust air, for
use in heating domestic hot water and for space heating. There is
rarely enough heat in the exhaust to permit the heat pump to carry
the whole heating load of the building, but the contribution is useful, and a heat pump can be very efficient when operating off a
heat source of 70°F.

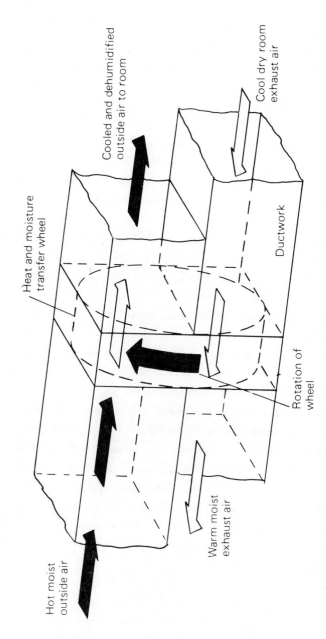

Figure 8-13. Enthalpy wheel—summer cooling.

Cooled and dehumidified outside air to room

Cool dry room exhaust air

Heat and moisture transfer wheel

Ductwork

Rotation of wheel

Hot moist outside air

Warm moist exhaust air

HVAC SYSTEMS

Now that the basic kinds of HVAC systems found in modern office
buildings have been described, we can begin to discuss measures
that may be taken to improve the performance of these systems through
system management and retrofit. The basic rule in improving HVAC
efficiency is to avoid situations in which heating and cooling occur at the same time in the same space. Instead of merely providing
a way to achieve a level of conditioning that is sufficient for
most commercial buildings, we wish to seek ways to "fine-tune" the
HVAC to a set of space demands that have, one hopes, been reduced
through the measures mentioned in the preceding section of this
chapter.

Some conservation measures apply to HVAC systems regardless of
the type of system in the building. Air-side ducting should be
inspected, and all impediments to smooth airflow removed. The objective is to have ducting that operates with the lowest possible
pressure drop. At the same time, since almost all existing HVAC
systems are oversized, it will surely be possible to reduce the
flow rate of supply air. Since reductions in fan volumes result in
more than proportional fan power reductions, savings on the order
of 4% are possible from this measure alone (4, pp. iv-15).

It is generally desirable to install variable air volume (VAV)
controls in the HVAC system: the VAV concept is more energy-efficient than any system employing constant air volume. Installation of
VAV boxes can be especially effective in buildings that have wide
daily load variation (e.g., major east and west elevations, large
glass areas, large surface area); annual savings, relative to constant air volume systems, can reach 50% in these situations. The
installation of VAV controls makes installation of variable-speed
supply fans cost-effective as well.

Any building with a three-pipe water system would benefit from
having a two- or four-pipe system. Significant economies are possible if a four-pipe system can be operated as a two-pipe system,
with only heating or only cooling being provided at any given time.

In addition to eliminating the possibility that heating and cooling will occur at the same time, the change to a two-pipe operation allows the removal of one of the two fan coils in each of the supply air ducts which in turn reduces resistance to airflow. Finally, it would be more efficient to be able to close off ducts that serve unoccupied areas of the building, for by doing so, energy can be saved on weekends, when only a few people may be in the building.

In a single duct, constant air volume system, the main objective should be to reduce airflow rates and fan power. After this, the next step would be to lower the temperature of supply air during the heating operation, if this could be done without affecting the ability of the system to meet peak load. This saves less energy than lowering airflow rates, but it is still important. Thus, one should find the optimal combination between airflow rate and supply air temperature. A similar trade-off exists in the cooling operation, but the efficiency of the central cooling plant is more dependent on water output temperature than is the efficiency of the heating plant. Reducing airflow saves energy; however, even more might be saved by raising the temperature of cooling water, thereby improving the coefficient of performance (COP) of the chiller.

System control in constant air volume systems should include an 8 to $10°F$ dead zone between the heating and cooling modes, to prevent rapid cycling and simultaneous heating and cooling in adjacent zones. VAV controls should be installed wherever possible. In the classic VAV system, the temperature of supply air is held constant, and loads are met by admitting varying amounts of air to the space. A great deal of energy can be saved by making supply air temperature a function of the position of the VAV dampers. Supply air temperature should be controlled so that the zone with the largest heating or cooling demand has its VAV control wide open. To achieve this, additional controls are also needed to vary the amount of hot and cold water delivered to the fan coils by the central plant.

If the VAV system does have reheat coils, they should be prevented from activating until after the VAV damper has reached its

minimum opening setting. In general, VAV systems that bleed unneeded
supply air directly into the return air duct are less energy-efficient than systems with variable fan volume. This is primarily a result of mixing cold air in the return with hot air coming from the
conditioned space.

HVAC systems with terminal reheat coils are inherently energy-
inefficient, but even if the reheat coils cannot be eliminated, much
can still be done to limit the waste. Most terminal reheat systems
operate with a fixed-temperature output from the cooling coil. This
temperature, often about $55^{O}F$, may be lower than the cooling requirement of most of the zones, which means that terminal reheat will be
required in those zones. A significant amount of energy can be
saved if the output of the cooling coil is made dependent on the
number of zones requiring reheat. The cooling coil output temperature should be allowed to rise until only about 10% of the reheat
coils are on. Even more energy can be saved by adding VAV boxes at
each terminal and by controlling them by thermostats in the corresponding zones. As cooling load drops, the HVAC system should respond by closing the VAV box, not by reheating, until the VAV box is
at least half closed. Only when the VAV box is closed to its minimum opening should reheating be allowed.

Multizone and dual duct systems are both based on the mixing
of hot and cold air to fine-tune the air input to each zone. Energy
can be saved if the extent of this mixing is reduced. The hot and
cold decks (ducts in a dual duct system) are usually set at temperatures far enough apart so that the peak cooling and heating loads
in any zone can be met. But most of the time the building is not
under peak load, which means that hot and cold mixing must occur in
all zones. Automatic temperature controls should be installed, to
raise the temperature of the cold deck (or duct) and lower the temperature of the hot deck (or duct), until the load in the zone with
the biggest cooling or heating load can just be met.

A multizone system is much more efficient if it has individual
heating and cooling coils for each zone. Such systems are now being

installed in many new buildings. The heating and cooling temperatures can be set individually for each zone, which greatly reduces
the extent of hot and cold mixing, and if these temperatures can be
made to follow the space load, or if VAV boxes are introduced, the
hot and cold mixing can be reduced even further. The large savings
from use of individual coils may even justify the replacement of an
existing multizone system with a new system which has individual
coils. The efficiency of multizone systems can sometimes be improved by converting zones that experience little temperature variation to VAV operation.

In induction systems, primary air temperatures for heating and
cooling are usually fixed, and the secondary water coils are used
to adjust the temperature in each zone. Primary air temperatures
should be modulated as functions of general load levels, and secondary heating and cooling reduced to a minimum. It may also be possible to reduce the primary air volume and pressure without impairing the system's ability to meet peak loads. Then secondary water-
flow rates can perhaps be reduced as well.

A very important way to save energy in many climates is to make
use of an economizer cycle for cooling. The economizer cycle is
nothing more than the use of outside air to cool the building, when
weather conditions permit. It is similar to the use of outside
ventilation air to cool a building at night and is particularly effective in dry climates with less than 8000 wet bulb degree-days a
year.

An economizer cycle can save energy whenever the total heat
content, or enthalpy, of the outside air is less than that of the
air in the conditioned space. Most of the time this will be true
when the outside air is $5^{\circ}F$ cooler than the inside of the building,
and this is the simplest way of deciding when to change over from
chiller to economizer cooling. If the outside dry bulb temperature
is lower than the return air temperature but too high to cover the
entire cooling load, the economizer can still relieve part of the
cooling load. In humid climates, with more than 8000 wet bulb
degree-days a year, it makes sense to control the economizer with

wet bulb or enthalpy sensors. Outside air is introduced for cooling purposes whenever its total energy content, or enthalpy, is
less than that of air being recirculated in the building.

Fred Dubin reports an instance in Denver in which an economizer
cycle saved 35% of the energy required for cooling but it is important to note that economizers do have disadvantages. In dry
climates, the introduction of significant amounts of outside air can
impose significant humidity loads that may cancel some of the free
cooling gains. Also, most economizer systems require the installation of a return air fan to move the additional quantities of air
through the ducting, so that new ductwork may even be required.
Finally, economizer systems tend to reduce the attractiveness of
heat recovery systems.

CENTRAL HVAC PLANT

The central HVAC plant consists of a boiler for heating and a chiller for cooling. A central chiller is usually on the order of 10
to 15% more efficient than a chiller in a packaged plant, and the
same is almost certainly true for boilers.

It is very likely that the central plant in the typical office
building was originally sized too large, and this oversizing will
be made worse by any measures taken to improve the efficiency of
either the building shell or the HVAC system. The central plant is
rarely as efficient at part load as it is at full load, but temporary adjustments can be made until permanent plant changes are
undertaken. For instance, smaller burners can be installed in
boilers, and the boiler can be tuned to operate efficiently at lower
output and with a minimum of excess air. Flue gas temperature and
composition should be monitored. Heat recovery schemes should be
investigated, so that waste heat from the boiler flue can be recovered for use in preheating boiler makeup water, combustion air,
or sanitary hot water. Boiler blowdown should be performed as
needed, and not on a fixed schedule. If the boiler plant consists
of multiple units, units should be loaded to capacity before addi-

tional units are brought on line, since one boiler at full capacity
is much more efficient than two at part load.

The same general concerns apply to the chiller plant. However,
in contrast to the boiler, the temperature of the chiller's output
water has a strong influence on its COP; thus, the output temperature should be chosen in concert with other changes being made to
the HVAC system. It may be possible to choose higher output temperatures, improve chiller efficiency, and still cover peak cooling
loads. In general, the evaporator temperature should be raised, so
that at least one coil control valve is open all the way, even though
additional pump and fan energy would be required to circulate larger volumes of lower temperature water and air. As with boilers,
a plant of multiple units should be managed so that the overall
efficiency is optimized at every state of load. Off-line chillers
should be isolated. Energy may also be saved by connecting chillers
in series instead of in parallel.

A heat pump is an attractive choice in buildings whose heating
and cooling loads are well balanced. However, the installation of
a heat pump is likely to be complex, because the heating and cool--
ing loads will rarely be matched. The heat that may be extracted
from the cooling water may not suffice to cover the heating demand.
In this case, more feed water must be added to the cooling loop to
prevent its temperatures from dropping too far, or a backup heating
system, either electric resistance heater or boiler, must be installed. If the cooling demand exceeds the heating demand, a cooling tower may be needed to vent excess heat. Another problem is
that heat pumps have difficulty delivering heat at the elevated
temperatures (80 to $100°F$) required by fan coils in conventional
HVAC systems.

Condenser efficiency can be improved by installing cooling
towers. The condenser temperature should also be varied according
to the load requirement, so that the cooling tower runs continually,
instead of cycling on and off. Lowering condenser temperatures can
improve chiller COP from 10 to 40%, depending on the type of chiller.

Cooling towers must be kept clean to be efficient, and it may also be desirable to consider cold water storage to allow the chiller to precool water at night, when the outside air is coolest and the cooling tower is most efficient.

OPERATIONS AND MAINTENANCE

Current HVAC systems are inefficient in design and operation, and they are also poorly maintained. They abound with leaks, dirt, and equipment that should be replaced. Implicit in all the conservation measures discussed is the need for a much finer and more thorough maintenance of the HVAC and of the entire building than is now usually the case. Otherwise, much of the value of more sophisticated design and operation will be lost.

Although all the areas of building performance in which maintenance is crucial cannot be covered, a few maintenance-intensive aspects of conservation in commercial buildings should be mentioned:

1. *Lighting.* Unnecessary lamps should be removed from overhead fixtures, and the remainder maintained to conserve their efficiency.
2. *Infiltration.* Combatting air leakage into and out of the building envelope is a never-ending task. It is certainly worthwhile to seal cracks, install weather stripping, and check outside air dampers on windows and doors leading to the outside or to unconditioned spaces.
3. *Central plant tune-ups.* Boiler tune-ups and cleaning are cost-effective measures for improving building efficiency.
4. *HVAC ducting.* Parasitic losses through cracks and leaks in ducting can seriously impair the efficiency of HVAC systems. Regular checks should be made to ensure that duct joints are tight and, at the same time, additional insulation may be installed.

5. *Managing the building.* Operating a building for maximum efficiency is a daily job. Maintenance supervisors can contribute by inspecting the building, turning off unnecessary lights and equipment, closing draperies and shutters, turning down conditioning in unused spaces, and the like.

In short, the building must be aggressively maintained and managed if the full value of the conservation measures is to be realized. It may be appropriate to make the maintenance supervisor's annual bonus dependent on his or her performance, relative to the previous year, in helping to save energy.

BUILDING MANAGEMENT AND CONTROL

The conservation measures described above imply and require sophisticated control of the HVAC system. Temperature levels and flow rates for water and air systems must be changed continuously, to allow the system to follow the building load as it changes during the day. Equipment must be shut down at night and started up in the morning, and ventilation rates and economizer cooling must be controlled. Finally, the central plant must be managed and maintained.

A central control system can assist the maintenance staff in managing, in an optimal manner, the building and its many systems. Though such a system must be custom-designed, a central control system is usually cost-effective for a large building (above 100,000 ft^2) and can achieve a 20% savings as well as pay back its cost in 3 years.

A diagram of a typical central control system is shown in Fig. 8-14. The system optimizes building operation, according to data received from remote sensors. Some of the sensors are yes-no devices or alarms, and some are analog readouts. In addition to controlling HVAC operation, the system can manage electrical demand

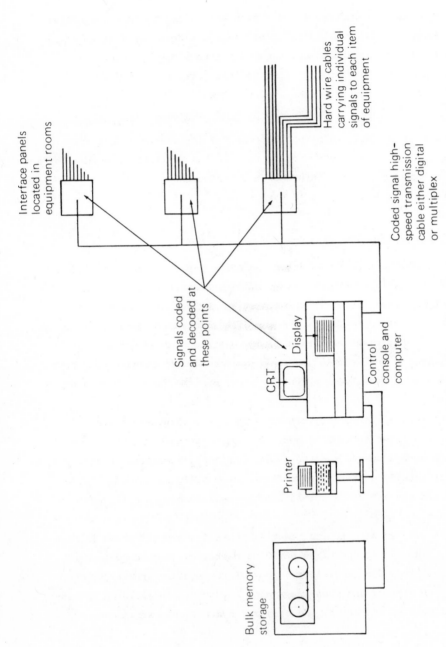

Figure 8-14. Typical central control system. (From F. Dubin, H. Mindell, and D. Bloome, Guidelines for Saving Energy in Existing Buildings: Engineers, Architects and Operators Manual, p. 307.)

to limit demand charges, signal operators if any equipment exceeds operating limits, and even provide security surveillance. It can collect data on energy consumption and schedule maintenance. A listing of some typical programs and their interfaces is shown in Fig. 8-15.

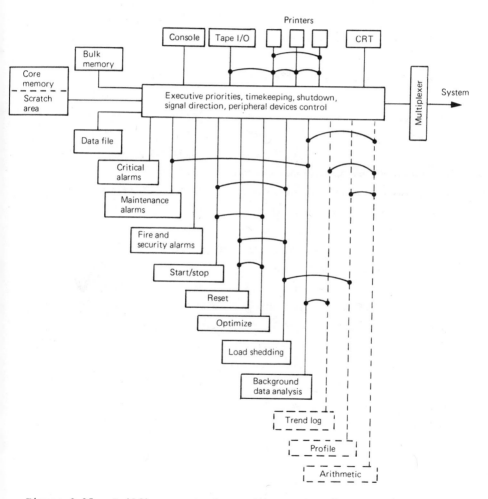

Figure 8-15. Buildings central system--programs and interface. (From F. Dubin, H. Mindell, and D. Bloome, Guidelines for Saving Energy in Existing Buildings: Engineers, Architects, and Operators Manual, p. 313.)

BUILDING ANALYSIS TECHNIQUES

There are no standard formulas for making office buildings more
energy-efficient. Each building is a unique design with a unique
microclimate and must be analyzed as a special case. Certain measures, such as reducing infiltration or avoiding reheat in HVAC systems, are obvious and can be recommended without hesitation. But
most conservation measures will influence more than one aspect of
building operation. The effect of any one of them may be difficult
to estimate, and the effect of multiple measures taken on the same
building even more so.

In new construction, the complexities of system operation have
led engineers to use conservative rules of thumb in sizing equipment
and in specifying operating procedures for the HVAC system. We
have seen how critical the temperature settings of air supply can
be. In the future, more efficient commercial buildings will require
yet more complex operational procedures, and one can reasonably
wonder whether simple rules will still exist for fine-tuning the
operation of an office building.

The difficulty of understanding the energy operation of commercial buildings has been recognized for years, and a whole industry
that specializes in modeling the energy behavior of office buildings
has developed. In the United States, undisputably the world leader
in this sort of modeling, there are least *10* to *15* companies that
specialize in computer simulation of office buildings.

Most of the present generation of American analysis programs
are based on programs developed under federal sponsorship in the
1960s: Many are descended from the "Post Office" program developed
for NASA. These analysis programs were first developed to help
designers size HVAC systems in new buildings, but many of them are
now being used to assess the attractiveness of retrofits to old
buildings. Designers are adding more and more complex programs to
handle more and more building types and situations. The overall
logic of these simulation programs is the same: Each one begins
with loads on the conditioned space and moves through the HVAC

system to the central plant and, finally, to energy consumption.
The programs consider first the varying loads on the conditioned
space, including gains from thermal transfers through the walls,
solar gain, lighting, and people. Many of the programs simulate
the heat transfer behavior of the building in an extremely sophisticated manner. Others use constant factors derived from the well-
known fundamentals of ASHRAE. The programs even consider the transfer of radiant solar gain into thermal loads in the space. The programs then simulate the response of the HVAC to these loads. The
response depends, of course, on both the system and its control procedures. Finally, the programs simulate operation of the central
plant, modeling changes in efficiency as the plant follows the loads
on the air distribution systems which serve the space. The final
outputs are annual (or daily) energy use and peak heating and cooling loads.

The ability of these programs to reproduce the actual behavior
of an office building depends on how well they can simulate reality.
It also depends on how accurately the building operation is specified
and on how the analyst interprets the specifications. A number of
major projects to compare the results of such programs when applied
to the same building have been performed. Results have shown that
the interpretations that analysts place on the specifications can
change the answers as much as can real differences between their
programs. In one such project carried out in Canada, a number of
different analysts produced widely differing results in applying
the same analysis program to the same building specification. Because some divergences damp out over longer runs, differences between programs are more marked for daily energy consumption than they
are for annual energy use.

Such results question the ability of analysis programs, however sophisticated, to reproduce actual building performance. Most
of the error is likely to be in the specification, which may not
reflect true building operation or occupancy patterns, or in its
interpretation, since these are not likely to change if the same

analyst considers a number of modifications to a building. However, these analysis programs can still be very useful to building owners who wish to improve the performance of their buildings. Indeed, they seem to be the only way to consider all the variable measures and their effects at once.

NOTES AND REFERENCES

1. Fred S. Dubin, New energy conservation ideas for existing and new buildings, *Specifying Engineer,* Cahners, Boston, January 1976, pp. 65-72.

2. Jerry Jackson and William S. Johnson, *Commercial Energy Use: A Disaggregation by Fuel, Building Type, and End Use,* ORNL/CON-14, Oak Ridge National Laboratory, Oak Ridge, Tenn., February 1978.

3. In Minneapolis, there is 25% more sun on a south window and 15% more on an east or west window than there is on a north window. In Florida, the amount of sun striking all windows is about equal.

4. Mathematica, Inc., *Comprehensive U.S. Evaluation of Energy Conservation Measures,* EPA 230/1 75 003, Environmental Protection Agency, March 1975, pp. iv-22.

5. John K. Henderson, Fundamentals of air conditioning, in *Engineering and Building Structures,* vol. II, LeRoy E. Varner, Sr., ed., Building Owners and Managers Institute International, Rosemont, Pa., 1973, pp. 9-34. A ton of air conditioning is equivalent to the removal of a load of 12,000 Btu/hr from a space.

6. Enviro-Management and Research, *Promotional Information for Energy Conservation,* U.S. Department of Energy, March 1978.

9

The Role of the Federal Government

Maxine L. Savitz
U.S. Department of Energy
Washington, D.C.

We are, as the historian David Potter once called us, a "people of
plenty." Over the past 250 years, the nation has grown accustomed
to an abundance of natural resources that has provided us with ample
opportunities to experiment with the traditional building materials—
wood, clay, and stone. The gradual development of new technologies
introduced such additional materials as brick, metal, cement, and
glass, but it was only after World War II that a whole new series
of building materials was developed, a series in which each new material was a substitute of a synthetic for a natural material:
vinyl asbestos flooring for asphalt tile; fiberglass and polyester
fabrics for cotton and wool; plastic laminate surfaces for natural
plywood. And in each case the substitution required a substantial
increase in the amount of energy needed to produce the new material (1).

Nature bequeathed the early settlers a rich legacy. At a time
when parts of Europe had already experienced serious deforestation
(by the early seventeenth century, for example, the Venetian Repub-
lic was forced to have its ships built in Northern Europe, because
its indigenous lumber supply was completely depleted), the New

England colonists were confronted with an apparently limitless ex-
panse of virgin timber. Using this readily accessible building material, the settlers adapted the traditional wood skeleton and plas-
ter structures they had brought from Europe to the more rigorous
climate by adding insulating layers of wooden clapboard on shingles.
As their buildings became more sophisticated, they compartmentalized
the rooms along central hallways, using doors to close off these
rooms during the winter months in order to conserve heat. In the
Deep South, the available building materials were essentially the
same as those in the Northeast, but the subtropical climate obviously presented different problems. There, houses were raised
above flood level and out of reach of animal and insect pests. This
elevation offered maximum exposure to prevailing breezes. Huge,
parasol-type roofs and continuous porches and balconies protected
the house against sun and rain, while tall ceilings, central halls,
floor-to-ceiling doors, and windows with louvered jalousies provided
maximum ventilation. In the Midwest and Southwest, the earliest
settlers also utilized the most readily available materials to build
their shelters--in the former area, sod houses, in the latter, adobe.
In each case, the types of buildings in the region were designed to
accommodate, not to confront, the exterior environment.

By the latter part of the eighteenth century, however, there
were already indications of the breakdown of these regional building
patterns. The invention of the steam-powered sawmill at the end of
the eighteenth century and the development of machine-made nails in
the 1830s contributed to the growth of a giant lumber industry on
the eastern seaboard. New England began exporting both lumber and
prefabricated building materials as far away as the Midwest. Three
other technological innovations aided in the homogenizing of regional architecture: the development of the metal structural skeleton; the invention of rolled glass in the 1880s, and the use of gas,
oil, coal, and eventually, electricity to heat, illuminate and, more
recently, cool building interiors.

Paradoxically, these technological developments over the past
150 years proved to be both blessing and curse. They made possible

the multistory building and the skyscraper; they freed up interior space, since heating and cooling systems no longer depended on a series of boxed-up rooms strung along a central hallway. At the same time, however, this technological progress, predicated in part upon a continuous supply of cheap and abundant energy, brought into being a massive, energy-inefficient building stock.

Of all the technological changes that have influenced the design of today's buildings, cheap sources of energy for heating, cooling, and illumination perhaps have had the greatest influence on the buildings sector. The following, written by a historian of architecture in the early 1960s, neatly sums up the premises under which most of our current buildings were designed:

> Protection from sunlight, heat and cold, has, over long periods of time, been more important than security. Wars and civil disturbances come and go; climate changes slowly, and the unending battle with weather has inspired inventions that have enriched architecture. The covered colonnade, for example, gives protection from the direct rays of a strong sun; the sloped or pitched roof deals effectively with rain and snow; and the vault and the thick wall have, in the past, defeated extremes of heat and cold, just as air-conditioning and central heating can now secure comfort in any climate. (2)

This emphasis on the comfort of the interior environment to the exclusion of any consideration of exterior conditions, either of siting or climate, and the assumption that cheap energy will permit these aspects of building design to be ignored, constitute a powerful mental legacy. It is a legacy which has profoundly shaped our current stock of residental and nonresidential buildings.

THE NATURE OF OUR CURRENT BUILDING STOCK

A brief look at the energy use problems presented by one type of building will illustrate some of the difficulties we face in introducing energy conservation measures into our enormous (75.5 million dwelling units) and slow-to-change stock of existing buildings. The glass-walled office or apartment building, whether several storied or multistoried, is a predominant building type throughout

the country, but it has seldom been designed or operated with any
regard to the siting requirements of different exposures and different climates. We dwell in a country with enormous regional variations in climate, yet north and south walls of these buildings,
whether in Minneapolis or Phoenix, are identical in design. Consider what this means in terms of the energy burden placed on the
heating and cooling units of such a building:

> On a cold, bright, windy day in December, the north wall--
> chilled by the wind and untouched by the sun--would have the
> climate of Canada. At the same time, the south wall of the
> same building, protected from the wind and exposed to the sun,
> would have a climate like that of South Carolina on a hot July
> afternoon, while at the same time, the east wall would have the
> climate of Massachusetts. (3)

Such buildings are the rule and not the exception. Furthermore, when we consider that about 37% of the nation's current yearly
energy production, that is, about one-sixth of the nation's oil and
one-fourth of its natural gas, plus a large portion of its peak
electrical load, is used by its buildings, we can begin to understand one dimension of the problem facing us.

Another dimension is the diffuse and fragmented nature of the
buildings sector. Traditionally, all matters having to do with
construction and building are supervised by local governments; the
result is thousands of different building codes and standards. Compared to this sector, the automobile industry appears almost mono-
lithic--if one wants to create a more energy-efficient car, Detroit
is the obvious place to begin. The buildings sector has no such
immediately identifiable locus: It is a composite of hundreds of
thousands of homeowners and renters, manufacturers, public services,
suppliers and builders, which is supported by a host of financial
institutions and regulated by a myriad of different municipal,
county, and state laws.

THE TRADITIONAL FEDERAL ROLE IN ENERGY

Any discussion of the nature of the federal role in the conservation of energy in the buildings sector must consider the interpretation of such terms as "the common good" or "the public interest" as well as the way in which we have interpreted them in the past, for what becomes immediately evident is that every person in this country is affected to some degree by what happens in the buildings sector. Moreover, the pervasive impact of the energy crisis on every segment of our society has placed the Federal government in the role of mediator and broker among the various conflicting interest groups, simply and primarily because it is the only institution with the resources—and the legitimacy—to act as a unifying force.

Historically, the federal role in energy-related matters had been to act as a conservator of national resources by managing national forests, mineral reserves, and reclamation projects, and by establishing a network of regulatory agencies and commissions as a response to specific, limited, and immediate problems. The government has also acted as

> A promoter and developer of energy technologies and systems such as hydroelectricity, oil pipelines, and nuclear power. Furthermore, the government has substantially aided energy users in the transportation, manufacturing, farming, and domestic sectors of the economy. Prior to the establishment of the Department of Energy, however, none of this extensive federal activity was subject to coordinated planning and management. (4)

Except in times of extreme national emergency, such as war, the federal government has been reluctant to intervene directly in the management of our national energy resources. For example, although the government was a direct participant in the development of our hydroelectric power resources, it was only after the problem of conflicting interests became evident that the government began, prior to World War II, to supervise the national use of petroleum. How-

ever, it was not until competition among various groups and regions
began to threaten the common good that the government directly
intervened to set up the Petroleum Allocation Board. In this case,
neither volunteerism nor the hitherto established federal practice
of leaving the issues of long-range planning in energy matters to
private industry or local and regional authorities worked; the
government was forced to step in. (5)

THE FEDERAL ROLE TODAY

The United States has prospered for nearly 100 years because the
availability and price of energy were not pressing concerns to
Americans. We developed casual and wasteful ways toward our nonrenewable energy resources, which were inexpensive and seemingly
inexhaustible. Our natural appetite for oil rapidly exceeded our
ability to product it. The United States--along with most of the
world's industrialized nations--turned increasingly to a small group
of countries for vital oil supplies. By 1973, the United States depended on those few nations for 40% of its oil supply, and other
countries were even more dependent than we.

Several events in the 6 years since 1973 have provided painful
reminders of the folly of this almost total dependence on depletable
fossil fuels and of this overreliance on imported energy:

The oil embargo of 1973 and 1974
The natural gas shortage in the harsh winter of 1976-1977
The coal strike in 1978, which significantly reduced electric power
    resource margins
The recent reduction in Iranian oil imports

The rising price of energy, relative to other factors in our
economy, requires consumption adjustments, in order to ensure that
a healthy economy and full employment can still be achieved. Energy
in general, and oil in particular, are no longer cheap. We must
find ways to cut our consumption and still fulfill our needs: This
is the essence and the purpose of energy conservation. The extent

to which demand for energy can be reduced, without substantial effects on economic prosperity and growth, is a major issue, and the implementation of cost-effective conservation measures has the potential for resolving that issue. Conservation does *not* imply significant sacrifice, discomfort, and a reduction in the standard of living. In fact, conserved "energy" can be thought of as a source of energy supply, since displacing energy at the final point of use is more effective and usually less costly than producing an additional unit of energy at the source.

In 1978, the U.S. gross national product (GNP) was $2.1 trillion, and the country consumed 78 quads of energy. The United States imported an average of 8 million barrels of oil per day equivalent, which is 22% of its energy supply and 44% of its petroleum supply. In monetary terms, the energy cost $89 billion, and despite the fact that this comprises only 4.2% of the GNP, rising energy costs and the growing, increasing inflation, weakening the dollar abroad, constraining our foreign policy options, and threatening our national security. Also, sporadic supply shortfalls have caused enormous national discomfort in terms of gasoline station lines, lack of adequate heating fuel, disruption of business, and other problems. In an attempt to alleviate our dependence on foreign oil, in October 1979, President Carter established the national objective of steadily decreasing oil imports to 4.5 million barrels per day by 1990. The U.S. government has also responded with a number of policies and programs designed to both increase production and encourage a reduction in demand.

The U.S. program for energy conservation is based on a number of premises (9):

1. Through deregulation of certain oil products and natural gas, energy prices should reflect the true replacement cost of energy supplies. This will move the United States closer to a marketplace in which advanced conservation technologies—and more efficient energy usage—will become economically attractive.

2. Government-sponsored information/demonstration/commercialization programs, which have been developed and are being implemented, will help the marketplace work as effectively as possible.

3. Government-sponsored RD&D makes more options available to the market sooner by spreading development risks, augmenting limited private sector RD&D funds, and improving the economics of various technologies.

4. Because of the national importance of improved efficiency in energy-related capital stocks, existing market mechanisms are being augmented by economic incentives/penalties to help overcome the reluctance of business and consumers to invest in energy efficiency.

5. Because of the national importance and urgency of reduced petroleum use and of improved end use efficiency, mandatory regulations and standards are being promulgated to ensure significant market penetration of alternate fuels and conservation technologies.

To date, substantial progress had been made in energy conservation. For the period 1950-1973, the GNP and energy demand grew at virtually the same annual rate--about 3½%. Since 1973 and the Arab oil embargo, energy demand grew at just under 1% per year, while the GNP grew at over twice that rate (2.3%), and while the economy continued to grow, energy demand in the industrial sector actually declined at about 1.2% a year (see Table 9-1).

Table 9-1. Annual GNP and Energy Consumption Growth Rates (Percentage per Year Average)

Sector	1950-1973	1973-1978	1977-1978
Residential and Commercial	4.1	+ 2.6	+ 3.2
Industrial	3.1	− 1.2	+ 0.1
Transportation	3.3	+ 1.7	+ 2.6
Total Energy Consumption	+3.5	+ 0.9	+ 1.9
GNP Growth (1972 $)	+3.7	+ 2.3	+ 4.4

Although the United States has been a prodigious energy consumer, in relation to other countries' energy growth rates over the
past 20 years our record is not too bad. As shown in Table 9-2,
between 1960 and 1976, our historical energy growth rate has been
lower than that of most of the developed world. Our projected energy growth is also comparable to that of the developed world.

Moreover, for the last half-century in this country, the amount
of energy required per unit of GNP has generally declined, despite
the declining real cost of energy during that period. In 1947, for
example, 115,600 Btu were needed per dollar of GNP. By 1960, this
figure was reduced to 92,7000, and by 1976, only 79,800 Btu were
associated with each dollar of GNP. Thus, recent historical trends
provide evidence that the United States is indeed reducing its energy
appetite and that continued economic growth without equivalent growth
in energy use is possible.

Conservation is the collective result of decisions made by
millions of individuals and businesses in their everyday affairs.
Federal government activities are only one factor influencing these
decisions; state and local government policies also have direct impacts. Six general principles prior to 1981 shaped the federal
government's approach to conservation activities:

1. Conservation programs should be designed to reduce waste by
    finding more efficient ways to conduct present and future ac-
    tivities.

2. Conservation activities should be undertaken in ways that pre-
    serve and improve the quality of American life.

3. There should be a maximum reliance on the free market to allo-
    cate energy resources.

4. When financial incentives are used, they should enhance the
    present and future operation of the free market.

5. There should be a minimum of federal government regulation or
    direct interference with decision-making in the private sector.

6. The federal role in conservation is important but should not be
    predominant. Conservation techniques and objectives should be

Table 9-2. *World Energy Growth Rates: History and Projections, Series A-E, 1960 = 1995* (Percentage)

Supply Curve Demand Curve	1960-1975	1975–1995				
		A High High	B Low High	C Mid Mid	D High Low	E Low Low
Energy Growth Rates						
United States[1]	3.3	3.0	2.2	2.4	2.4	1.9
Canada	4.5	3.4	2.6	2.6	3.0	2.2
Japan	8.7	4.9	4.1	4.3	4.4	3.8
OECO Europe	3.9	2.8	2.2	2.2	2.3	1.7
Austria: New Zealand	5.0	3.7	2.7	2.7	2.7	1.5
Total OECO	4.0	3.2	2.4	2.5	2.7	2.1
OPEC	8.0	8.0	7.9	7.7	7.2	7.2
Other[2]	8.3	5.6	5.0	5.1	5.2	4.6
Total World	4.3	3.9	3.2	3.2	3.3	2.8
Energy GOP Growth Ratios						
United States	0.93	0.78	0.50	0.70	0.30	0.87
Canada	0.88	0.85	0.67	0.73	0.91	0.68
Japan	1.08	0.80	0.70	0.75	0.82	0.72
OECC Europe	0.93	0.73	0.57	0.54	0.72	0.57
Australia: New Zealand	1.32	1.02	0.77	0.84	0.93	0.55
Total OECO	0.38	0.74	0.58	0.58	0.76	0.60
OPEC	0.96	1.06	1.03	1.06	1.06	1.06
Other[2]	1.31	0.96	0.87	0.93	1.00	0.93
Total World[2]	0.94	0.84	0.71	0.77	0.85	0.73

[1]Includes Puerto Rico and Virgin Islands.

[2]Excludes Communist countries.

tailored to the varying needs of regions, states, and local
areas, and state and local government participation should be
steadily increased in the allocation of resources, in the operation of government programs, and in decisions on alternatives
with respect to conservation.

Within these general guidelines, the opportunities to increase
the efficiency with which energy is used are enormously diverse.
Some opportunities are appropriate targets for federal activities,
because the federal government has a unique capability to act
throughout the nation. Other opportunities can be more effectively
addressed by state or local governments, private associations, businesses, or individual citizens.

Barriers to Conservation

Despite the significant potential savings from conservation measures,
there are many barriers that may keep the savings from being realized, and these barriers must be considered in the development of an
energy conservation policy.

Artificially Low Energy Prices. These are a major barrier to conservation. For years the American consumer has been protected from
the realities of the world market, because price controls on oil
and gas have held prices below their replacement costs. This has
encouraged inefficiency, since users have treated oil and gas as
cheaper than they really are. The current federal policy of complete
oil decontrol and accelerated deregulation of natural gas will dim-
inish this barrier over the next several years (6).

Imperfect Information. In order for the market to work well, purchasers must have enough information to make rational economic decisions, a situation that is often not the case with conservation.
Some individuals do not know the basic concepts of energy efficiency, and even knowledgeable individuals can have trouble discover-
ing the relative energy efficiencies of competing products or the
full initial costs of products when installation costs are involved.

Private Sector Investment Distortions. Ideally, decisions to purchase goods and services should be based on the appropriately discounted cost of an item throughout its useful life. Instead, consumers have a strong tendency to focus on the initial cost in deciding what to buy, and they tend to demand a relatively short pay-back
period. This causes a bias against conservation, because the costs
of conservation measures may accrue immediately, while benefits accrue slowly but steadily in the future. Moreover, because they lack
discretionary funds or access to credit, many households have little
opportunity to invest even in conservation measures with rapid returns.

Constraints on Private Sector Research and Development. The private
sector may not invest in energy-related research and development
at levels commensurate with the overall importance of conservation.
In many cases, energy research has a low probability of success,
and the value of any results depends on uncertain future energy
prices. Firms would be reluctant to spend resources for research
and development on energy if they believed their competitors were
likely to share any benefits without comparable investments. Private sector research is often geared to new products and not to
producing existing ones in a more energy-efficient manner.

Legal and Regulatory Constraints. Other barriers to conservation
can be described generally as legal and institutional problems; many
of these are beyond the reach of the federal government. They include local building codes and zoning restrictions that prevent or
defer conservation-oriented improvements to buildings, rates for
electricity that decline as the quantity used increases, and restrictions on sale of electricity from cogeneration plants. Changes
in these factors are often difficult because federal, state, and
local laws and regulations must be modified.

Policy Evaluation Criteria

The development of criteria for the selection of any government
policy is no easy task, given the multiplicity of goals any given
federal program is to achieve. Energy conservation programs are no

exception. Consider first the measurement of the benefits of various energy conservation policies. Should benefits be measured by the improvement over time in the energy efficiency of the process (Btu's per unit of output) or in the absolute descrease in Btu use? If improvement in energy intensity is the criterion used, projects which save a large fraction of a minuscule amount would be elevated over those which save a smaller fraction of a large amount. Should the calculation of benefits distinguish between the types of fuel saved, giving projects which save oil and gas more credit than projects which save coal or electricity? If such credits are given, policies which save oil and gas but actually increase total Btu use can be elevated over projects which save Btu's. Should Btu savings be discounted to reflect the fact that current Btu savings are worth more than future saving? If so, what should the discount rate be?

There are also differences in opinion as to whether the government should play an active role, especially in funding technologies that promise near-term benefits. Arguments have been proposed on both sides of the issue. Proponents argue that infusion of federal funds are warranted, regardless of the benefits that might accrue to certain segments of the private sector, because of the high social benefits that would result from increased reliance on domestic energy supplies and reduced energy consumption. Opponents argue that social benefits might actually be less, because competitors might be dissuaded from pursuing research programs in areas where the government is subsidizing research, development, and demonstration (RD&D) activities.

Consistent with these concerns, a project evaluation system which considers these factors and maximizes public benefits is needed. The primary objective should be to select projects which provide high-energy benefits while simultaneously reducing the potential for counterproductive competition between the public and private sectors.

The first stage of any evaluation should be related to the potential energy benefits that would result from the successful commercial introduction of the technology. In many cases, energy-

conserving technologies would never be developed without federal
support. In some instances, the purpose of federal involvement is
to introduce the technology at an earlier date, since by accelerat-
ing its introduction into the marketplace, the corresponding energy
saving will be realized sooner than otherwise expected. Figure 9-1
illustrates a situation in which support by the federal government
permits the innovator to accelerate the research program and thereby
introduce the technology at an earlier date. The total energy savings accruing from the accelerated introduction of the product or
technology are represented by the total area under curve A, and the
incremental or marginal energy savings resulting solely from the
federal program are represented by the shaded area.

Projects which have large energy savings potential per unit or
which can achieve greater acceleration into the marketplace will receive higher ratings with this evaluation criteria. However, it is
also possible for projects which have smaller energy savings or
shorter acceleration periods, but which have higher market penetration potentials (total share of the available market) to receive

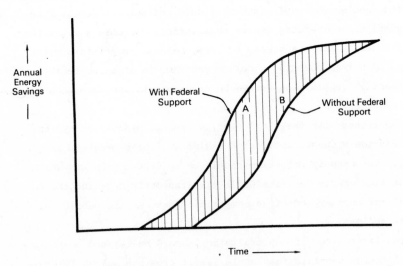

Figure 9-1. *Illustration of energy benefits.*

similarly high rankings. The fact that a project may offer high potential energy benefits or a more rapid acceleration into the marketplace does not make that project a viable candidate for federal funding, unless it can be demonstrated that energy benefits can be achieved through the efficient use of federal funds.

By subsidizing research, the federal government can reduce the overall cost of product development, can enhance product credibility, and can provide greater dissemination of the research results, all of which may tend to broaden market exposure and increase the rate of market penetration. Undoubtedly, these factors play an important role in stimulating the private sector to seek federal support for product innovations. However, these benefits could also exist for projects that the private sector might pursue independent of federal support. Therefore, in weighing the economic costs and benefits of projects seeking federal support, the problem becomes our ability to differentiate between those projects which the private sector should undertake without federal support and those for which federal support is required.

Figure 9-2 illustrates this problem. The vertical axis measures the potential net economic benefits to the innovator, and the horizontal axis measures the probability that the innovator will actually realize the benefits. Schedules A and B may be viewed as indifference curves. Projects falling on these lines are regarded as equally attractive by the innovator, that is, he or she is indifferent as to which to pursue. The difference between schedules A and B, which have been correspondingly labeled private and public sector criteria, relate to the propensity to accept risk. The public sector is normally willing to accept greater risk than the private sector because of the social benefits that can accrue when a project is successfully completed. A formulation which captures these relationships has been defined as:

$$r_{min} = r_{nom}(1 - P)^{1/a}$$

where

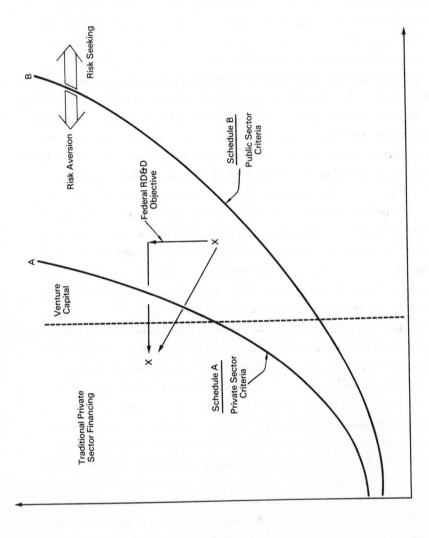

Figure 9-2. Benefit/risk trade-off criteria.

r_{min} = minimum acceptable return on investment

r_{nom} = nominal return on investment

P = probability of failure

a = propensity to accept risk (a > 0)

The propensity to accept risk is a function of the individual investor. The private sector trade-off schedule reflects risk avoidance (i.e., 0 < a < 1); the public sector schedule reflects risk acceptance (i.e., a = 1); and any schedule to the right of the public sector's reflects risk seeking (i.e., a > 1).

If these relationships hold, then there exists a region within which the Federal government can legitimately fund research projects, where the expected value is significantly greater than the public sector's social activities are small, and where large social benefits can accrue as a result of federal activities. The topology of projects which exist within this region may be characterized as having potentially high benefits to the private sector but as also exhibiting a high degree of uncertainty. The objective of the RD&D program from a commercialization perspective may therefore be illustrated as altering the position of a project from point X to X'. This may be accomplished by

1. Improving the potential net economic benefits
2. Reducing the risk of commercial success
3. Combining the two effects (as illustrated)

If such changes could be accomplished within the framework of an RD&D or a financial incentive program, the technology would then be in a position to compete for scarce investment capital.

Assessing the potential impact of federal support on the commercial viability of these technologies requires estimates of:

1. The potential return to the innovator
2. The risk perceived by the innovator
3. The propensity of the innovator to accept risk

If the research program or financial incentive is successful, federal support can have direct impact on the first two elements. If RD&D

expenditures are large, relative to manufacturing and marketing
costs, subsidization of the research program may improve the rate
of return sufficiently to justify commercialization. Likewise, to
the extent that the federal government conducts the research program,
the program should have positive impacts. The only factor that is
independent of federal support is the innovator's or, more broadly,
the private sector's willingness to accept risk.

A recent study stated "that only about 10 to 12 percent of the
ideas submitted for the initial screening and business analyses
process will enter the development pipeline which leads to commercialization" (7). This is an extremely low acceptance rate, which
implies a conservative investment philosophy within the private sector. It is even more conservative in view of the fact that of the
200 innovations that were studied--all of which failed after entering the "pipeline"--management still believed that 92, or roughly
half, were still good investments.

Andrew Sage has observed that the willingness to accept risk
is inversely related to the size of the capital commitment. This
would imply that the risk/rate-of-return criterion is probably also
a function of the capital at risk and is thus unique to every corporation (8). The extent to which this factor can be ignored in
the choice of private sector investment criteria selected for screening will depend on whether we prefer to err on the side of a strong
or weak justification for federal involvement. There are penalties
on either side. If a weak justification is selected, then the possibility of funding projects that the private sector should have
pursued, independent of public participation exists. If a relatively
stringent criterion is selected, it is possible that many promising
investments that the private sector will not pursue will also be
rejected for public support.

Policies for the Buildings Sector

The buildings sector is controlled by tens of millions of individual
owners and operators, each of whom has to make individual decisions
to carry out conservation measures. These decisions are influenced

by home builders, manufacturers, financial institutions, and state and local governments, as well as by the federal government. The technology for conservation is simple and generally available, but the imperfections in the market, particularly with respect to public information, make application of this technology difficult. In addition, capital stock in this sector turns over very slowly. For example, about 77% of existing residential buildings will still be standing in the year 2000, and new residential construction in a typical year adds only 3% of the total number of housing units. Furnaces, air conditioners, lighting equipment, and other energy-using appliances are usually replaced only when they stop working or become obsolete.

The building industry is highly fragmented and has accepted innovation slowly in the past. In many cases, builders are unfamiliar with existing conservation technology and its installation and are also reluctant to purchase new designs that add to the selling price of the building. Banks are reluctant to recognize the full value of conservation features when making loans, and few home builders or developers are large enough to do research and development (e.g., 90% of the home builders produce fewer than 25 units per year). The research and development that is done is concerned mostly with reducing production costs, not with improving energy conservation, and does not reach small builders.

The Arab oil embargo of 1973-74 stimulated for the first time significant interest at the federal level in energy conservation (10). The federal program supported a wide range of activities. In fiscal year 1981, approximately one-third of the Department of Energy's conservation budget focused on research and development, and demonstration. About one-half of the budget supported financial incentives or assistance such as weatherization for low-income people, energy audits and retrofits for schools and hospitals, grants to state and local governments to prepare and implement energy conservation plans. Only three percent went to activities related to congressionally mandated standards, regulation, and reporting, with

the balance used for education and information programs. Some of
these programs were expected to have immediate effects in saving
energy, e.g., weatherization of low-income homes, but research and
development is expected to produce results in future years.

The federal program had activities in all of the demand sectors
of the economy including the buildings sector. The basic objectives
of the federal buildings program developed in the 1970s were to reduce energy consumption in existing buildings by 20% between 1978
and 1985 and to reduce the energy used in new buildings by 30% (10).
The strategy for reaching these objectives was to work with the different constituents of the building industry (architects, engineers,
home builders, developers, community planners, manufacturers, etc.):
(1) to determine the energy used in a building for heating, cooling,
water heating, lighting; (2) to inventory and assess the research
that could lead to more efficient technology and to more efficient
building systems; and (3) to develop methods that would correlate
the commercial introduction and acceptance of new, more energy-
efficient practices and products. Many programs were congressionally mandated in this area (11).

It has been determined that such a comprehensive, aggressive
federal role in encouraging energy efficiency is no longer necessary (10). The most effective way of ensuring that the many
decision-makers from the homeowner to the corporate executive make
the decision to use energy efficiently is through deregulation of
petroleum and natural gas prices. Market price increases and ex-
pectation of increases in energy cost will accelerate the marketplace application of energy-saving technologies as well as the
development and demonstration of previously unproven technologies.
Thus, the federal buildings program in 1982 has been redirected
toward research in technologies and systems that are long-term and
broad based, involving high risk but offering large potential savings, but which are not yet attractive to funding by the private
sector.

From 1978 to 1981, the federal buildings program did make progress toward meeting the objectives established by Congress and the Department of Energy and its predecessor agencies. These results can be highlighted in the following areas:

Public Information. The free market operates most efficiently when each individual has all the information he or she needs to make rational economic decisions. Although the federal government had a number of information and education programs, these programs are no longer considered necessary as there is much information available from other sources. Some of the programs which were sponsored include:

1. Low Cost/No Cost Energy Conservation Information Program
2. Fuel Oil Marketing Program
3. Small Business Information Program

Financial Incentives. Financial incentives are provided for energy conservation through programs such as:

1. *Residential Tax Credit.* A 15% tax credit on an investment of up to $2000 in qualifying conservation equipment is now available. This investment must be made between April 20, 1977, and December 31, 1985, and must be for the improvement of a principal residence which has been in existence since April 20, 1977. Qualifying conservation equipment includes insulation, caulking, weather stripping, modified flue openings, storm or thermal doors and windows, automatic furnace ignition systems, energy-efficient replacement burners, and a meter which displays the cost of energy usage. This program is administered by the Department of the Treasury.

2. *Weatherization Assistance Program.* This program enables each state to purchase and install equipment to weatherize both owner-occupied and renter-occupied homes of low income families, particularly the elderly and handicapped. The list of approved equipment includes insulation, storm windows and doors, caulking and weather stripping, furnace efficiency modifications, and clock thermostats. To date, 758,000 low-income homes have been weatherized.

3. *Schools and Hospitals.* The National Energy Conservation and Policy Act (NECPA) of 1978 makes funds available on a 50% matching basis to individual schools and hospitals for retrofit investments. By the end of fiscal year 1981, more than 125,000 institutional buildings had been audited and 22,441 buildings had received grants for technical assistance and energy conservation measures.

Research and Development. Research and development projects were conducted in the areas of heating and cooling systems, home equipment design, lighting systems, energy efficient building shells, insulation, window design, ventilation requirements and systems, air infiltration characteristics, energy use estimating systems; and energy use control systems. In the space heating area, a highly efficient wood-fired boiler and nonsooting oil-fired boiler were developed and introduced for residential use. Both products used advanced combustion processes. Two oil burners using advanced atomization processes are scheduled for market introduction in 1982. These will increase system efficiencies by 15 to 20% over conventional design. In the appliance area, the development of the first successful heat-pump water heater (thereby creating a new industry) and an advanced refrigerator freezer were cost-shared. Research was performed for the first time on three-dimensional wall sections under static and dynamic conditions. Blower-door and tracer gas equipment to actually measure air infiltration into and out of residences were developed. The concept of "house-doctoring" for diagnosing comprehensive energy saving opportunities in homes was developed and tested.

Regulation. Since energy prices did not reflect the true cost of energy resources, Congress mandated several regulatory programs to achieve energy savings. Regulation, however, is not appropriate when price can achieve the same goal.

The following regulatory programs have been mandated.

1. *Building Energy Performance Standards.* This called for the development of nationwide performance energy-efficient design standards for new residential and commercial buildings. The effort has encouraged research that is widely applicable with or without regulation.

2. *Appliance Efficiency Standards.* The appliance efficiency standards were one regulatory element in a program that included the development of test procedures, support to the Federal Trade Commission in the appliance labeling program, and a conservation education activity. Labels are currently applied by manufacturers on major household appliances (e.g., water heaters, room air conditioners, refrigerators, freezers); thus, consumers are able to obtain reliable, comparative information when making purchase decisions. The standards portion of the program considers the need for mandatory nationwide efficiency for 13 product categories.

3. *Residential Conservation Service.* This legislation requires utilities to inform their customers about energy conservation measures. A utility must inspect the customer's residence, upon request, to determine where conservation measures would be most cost-effective. The utility must also provide lists of financial lenders, material suppliers, and contractors and offer to arrange with listed firms for the installation or financing of conservation measures. By late 1981, 47 states had developed plans for implementing this program and many utilities were already providing the conservation services.

4. *Federal Energy Management Program (FEMP).* The federal government is one of the largest consumers of energy in the United States. Its consumption accounts for 2.2% of our domestic energy use. The FEMP program involves developing guidelines for long-term building and operation planning, reviewing and approving the plan of federal departments, monitoring federal performance and making regular reports to Congress and the President. The federal government has reduced its energy use in buildings and facilities from

161 million barrels of oil equivalent in 1975 to 146 million barrels of oil equivalent in 1980--a 9.3% reduction.

Summary

As the government has implemented petroleum price decontrol and energy prices have increased, the Reagan Administration believes that it is no longer necessary for a comprehensive federal energy conservation program in the buildings sector. The thrust of the federal program will be in support of fundamental research activities. The private sector will undertake its traditional role of the development, testing and commercialization of near-term technologies.

CONCLUSION

Contrary to the perception of many, substantial progress in the energy conservation area has been made by the United States. Despite population increases and a rising gross national product (GNP), total annual energy consumption increased from about 76 quads in 1973 to barely 78 quads in 1980 (down from a peak of 80 quads in 1978 and 1979). This reduction translates into about 12 quads of foregone energy use, compared with 1973 efficiency levels; a savings of about $60 billion in 1980 energy purchases. The federal conservation efforts have played some role in this trend, although the exact effect is difficult to quantify--perhaps 5%. A recent report states that half of the reduction has been due to slower growth in economic activity. The remainder was due to improvement in energy efficiency, primarily as a result of using fuel prices and some as a result of government and utility conservation programs (12). As prices continue to rise and consumers become even more responsive and informed, energy efficiency improvement of buildings and its equipment should continue in future years.

NOTES AND REFERENCES

1. For instance, aluminum replaced steel as the product most often used for windows and door assemblies. Aluminum is an energy-intensive material; "on the average, its manufacture takes five times as much energy per pound as steel's."--Richard G. Stein, *Architecture and Energy*, Anchor Press, Garden City, N.Y., 1978, p. 51.

2. John Gloag, *Architecture*, New York, 1964, pp. 8-9.

3. James Marston Fitch, *Architecture and the Esthetics of Plenty*, Columbia Univ. Press, New York, 1961, p. 200. I am also indebted to Mr. Fitch for some of the information and the ideas discussed in the preceding section.

4. Richard Hewlitt, Foreword to In Quest of the Public Interest: The Federal Government as Managers and Guardians of Energy Resources, a session of the American Historical Association, Dallas, Tex., December 1977.

5. John A. DeNoro, The federal government as manager of petroleum resources, 1940-1942, paper presented at the American Historical Association, Dallas, Tex., December 1977.

6. Gradual decontrol of fuel prices is well underway. All rules and regulations concerning crude oil decontrol have been issued. Lower tier crude oil will be reclassified as upper tier crude oil by October 1981. Upper tier crude suppliers will be decontrolled at the rate of 4.6% per month through September 1981. Thus, by the time lower tier crude oil is reclassified as upper tier, controls will have been lifted from upper tier oil. The Department of Energy has recommended decontrol of heavy crude oil. Butane and natural gasoline will be decontrolled by Jan. 1, 1980, pursuant to a regulation issued on Nov. 29, 1979. The public comment period on a proposal to decontrol propane (liquified petroleum gases) ended Dec. 1, 1979.

7. Sumner, Myers, and Elden E. Sweezy, Why innovations falter and fail--a study of 200 cases, prepared for *Harvard Bus. Rev.*

8. Andrew P. Sage, Systems Analysis, manuscript pending publication.

9. As a result of the change in Administration and economic policies, there was a change in direction of energy policy which made a distinct break with the past. This Administration's energy policy has two pillars--production and conservation--and is based on a reliance on markets so that Americans have appropriate incentives to produce more and to use energy efficiently.

    The aims of the current Energy Conservation Research program are to conduct basic research on ways to make more efficient

use of available energy and to facilitate use of nonpetroleum
energy sources. In this context, conservation connotes using
energy more wisely while maintaining the Nation's standard of
living, rather than sacrificing and doing without.

10. Report to the Congress, Sunset Review, U.S. Department of
     Energy, February 1982.

11. The Federal Nonnuclear Energy Research and Development Act of
     1974 (P.L. 93-577) directed the Energy Research and Develop-
     ment Administration (ERDA) to initiate a national energy con-
     servation research, development, and demonstration program.
     The Energy Policy and Conservation Act of 1975 (P.L. 94-163)
     directed the newly created Federal Energy Administration (FEA)
     to develop test procedures and energy efficiency targets for
     major home appliances; and the legislation also established
     an energy conservation program for Federal buildings and opera-
     tions. The Energy Conservation and Production Act of 1976
     (P.L. 94-385) directed HUD to develop and promulgate building
     performance standards for new residential and commercial build-
     ings and mobile homes. The Department of Energy Organization
     Act of 1977 (P.L. 95-91) transferred these responsibilities to
     DOE. The National Energy Conservation Policy Act of 1978
     (P.L. 95-619) established the Residential Conservation Service
     Program, called for life-cycle cost procedures for Federal
     conservation investments in Federal buildings, and required
     consideration of minimum efficiency standards for 13 types of
     major home appliances. The Powerplant and Industrial Fuel Use
     Act of 1978 (P.L. 95-620) provided grants for areas "impacted"
     by coal and uranium mining. The Energy Security Act of 1980
     expanded RCS to include multifamily apartments and small com-
     mercial buildings, authorized auditor-training grants to
     states, and authorized the Residential Energy Efficiency Pro-
     gram, a building retrofit demonstration program.

12. Eric Hirst, *Energy Use from 1970 to 1980: The Role of Improved
     Energy Efficiency*. Oak Ridge National Laboratory, CON-79,
     December 1981.

Editor's Note: Most of this chapter was written in 1980, prior to
the election of the present administration and a distinct change in
direction of energy policy. See Footnote 9. for current role of the
Federal government.

10

Implications for the Future

John H. Gibbons
Office of Technology Assessment,
U.S. Congress, Washington, D.C.

One of today's most rapidly changing perceptions pertains to the manner in which we attempt to meet the needs that energy helps fulfill. After 200 years of seeking new supplies to sustain our increasing energy consumption, we have broadened our course and begun to take advantage of the cost savings opportunities provided by more effective use of energy. To our great good fortune, these new approaches have indicated that goods and services--amenities--can be provided by using much less energy than previously thought practicable or even possible. Improving the energy efficiency of building is one of the nation's best opportunities for energy conservation and our moving diligently in this direction can keep us off the rocky shoals of energy shortages and higher prices that otherwise lie before us.

Our buildings will undoubtedly be supplied with and utilize energy differently in the future, and the shape of that future will be determined by the alternatives we choose and, perhaps more important, by the rate at which we make the transition from old to new. As a matter of fact, moving rapidly or slowly may have a greater total impact on our economy and national security than the

final direction we choose. Our stock of buildings turns over at
only about 1% per year. While opportunities for greater energy efficiency are greatest in *new* buildings, we simply cannot focus all
our efforts on them. Much effort must also be expended on the more
difficult task of upgrading existing buildings. The time dimension
of world energy in the 1980s and 1990s makes it crystal clear that
we cannot afford, by any measure, to set a leisurely pace in improved energy productivity.

Prior to the 1973 embargo of Middle Eastern petroleum, the low
average prices of fossil fuel for heat and power, its apparently
endless availability, and our historical tradition of plenty meant
that the buildings market placed little or no value on tight building construction or efficient appliances. Keeping first costs low
and paying scant attention to operating costs was smart economic
sense.

No longer. Energy prices in the United States have begun their
move toward the much higher marginal cost that is now an unhappy but
permanent fact of life, and these prices are beginning to reflect
more accurately the real costs, since environmental and other previously "external" costs are increasingly borne by the user. Doub-
ling and redoubling market energy prices, with replacement costs
rising even faster, and uncertainty over reliability of supply from
old and new sources are now part of our national and personal awareness. Homeowners have made dramatic changes in rates of energy use,
going from sharp acceleration in the 1960s to stable per capita consumption in the 1970s (1), and disclosure, to propsective home
buyers, of past utility bills is becoming routine in many cities
and mandatory in others. Moreover, in response to market demand
and regulatory pressure, manufacturers and installers of energy-
consuming equipment are adjusting to this new climate of awareness.
Architects, engineers, and builders are coming to terms with the
factors that strongly affect our choices in design and construction.
The problem is *not* a lack in the number of productive responses to
higher energy prices; rather, the problem seems to be one of

accomodation to the kind of rapid change in which the time rate-of-
change of energy prices may leap ahead of traditional rates of turnover for energy-consuming capital goods.

While much has been learned about the technical improvements
that will increase energy productivity, less is clear regarding the
best methods for identifying and implementing the literally hundreds
of thousands of actions that collectively make up the conservation
response. We do know, however, about the importance of four elements:

Technical information, in appropriate detail in the hands of energy
    users
Capital
Trained people
Ingenuity in both technology and institutional arrangements

These elements are necessary for an accurate assessment of our
future energy potential, the implications of which may be examined
from three perspectives:

Future energy cost trends and availability
The opportunities that conservation presents for energy efficiency
    and resilience
The manner in which and the rate at which we should make the transi-
    tion from an economy of plenty to an economy of scarcity

ENERGY COST AND AVAILABILITY

It is clear that the pressures brought about by rising energy costs
and uncertainty about the availability of energy will help to pre-
cipitate measures to improve energy productivity in all types of
buildings. While industrial and commercial users, for whom the
energy cost is a large portion of their operating budget, may have
been the first to invest in conservation measures, even homeowners
are now responding to the market signals of rising average costs and
complementary public policies. Research conducted at the Office of
Technology Assessment in 1979 indicated a marked improvement in the

energy efficiency of new homes constructed in 1976, in comparison
with those constructed in 1973. These improvements in energy efficiency were achieved by wider use of simple techniques, with few basic
changes in building design, or without the use of more sophisticated
equipment. Interestingly, the level of thermal efficiency of the
1976 housing was already beyond the federally defined standards for
building efficiency that were only then coming into force in most
states.

In addition to builders themselves sensing a new market for
energy efficient buildings, homeowners and tenants have increased
their purchases of insulation, fireplace inserts, storm windows,
night setback thermostats, and other measures to hold down costs
and maintain comfort in existing housing. These responses show
sophisticated awareness regarding the tradeoffs between first cost
and operating costs. Also, the "minimum life-cycle cost" approach
is now being adopted by many public agencies and large purchasers,
among them the individual states and various agencies of the federal
government. Continuing investments made on this basis will not only
improve building efficiency but should lead to reduced unit cost of
energy-conserving equipment and materials through increased demand.

While curtailment in the supply of energy was the shock to our
national system which initiated the shift to conservation, continued
cost pressures and the uncertainty about future supplies and prices
are likelty to sustain the trend to improved energy efficiency and
result in the replacement of much of our major energy-consuming
equipment. The replacement of refrigerators, furnaces, boilers,
and entire heating, ventilating, and air conditioning systems with
models of vastly improved efficiency will sharply affect the baseload
requirements of utilities, especially throughout the coming decade.
Federal regulations that support these efficiency improvements will
place a much-needed threshold underneath the still-variable market
signals of energy price.

How far should we go, in terms of investing, toward making our
buildings less energy-intensive? The answer depends a great deal on

how high energy prices can be expected to go and, similarly, on how assured we can be about energy availability in the future. The real price of energy has shifted upward at a phenomonal rate since 1973, yet it is still no more expensive, relatively, than it was in 1950. The big difference between 1950 and 1980 is that, thanks to technology, we can now utilize other measures (e.g., the uses of insulation, electronic controls, and heat pumps) for enhancing energy efficiency in ways that simply weren't available to us in 1950.

The rise in the cost of energy cannot continue indefinitely. However, as prices move upward, more energy resources (in terms of both diversity and quantity) become competitive. The present marginal energy prices are probably well within a factor of 3 or 4 of their upper credible limit (in constant dollars). This kind of reasoning must be taken into account when we try to decide how much effort and how great an investment should be dedicated to energy efficiency.

We hazard a bet that the larger of two problems of energy supply--the rate of increase of the price of energy and its ultimate price--is the former. Because of the fact that we cannot respond quickly to higher energy prices--either through improved efficiency or (especially) through increased domestic supply--the likelihood is that real energy prices will continue to move up as sharply as possible, without inducing major international recessions. If this be the case, our priority should be to retrofit, as rapidly as possible, existing buildings to a *minimum* level of efficiency indicated by present *marginal* energy prices.

Clearly, the technology for conservation that is commercially available today is matched to far lower energy prices than now exist and are far removed from future energy prices. It should be self-evident, then, that a vigorous and sustained research program for the development of a basis for a conservation technology that is matched to future energy prices should be put into effect. This is not being done, despite an impressive number of potential new options (2-4).

Research in both the private and public sectors, in combination
with rising prices, will provide entirely new approaches toward
satisfying the energy-related needs of comfortable habitation. New
design concepts will dramatically lower the demand for space conditioning. The increased use of on-site energy sources is certain.
Most of the housing constructed throughout the decade of the 1980s
can be expected to resemble housing as it was constructed in the
past, but the new housing will be much tighter and more efficient
and will contain more sophisticated electronic sensors and controls.
It will have a greater ability to both retain and block out solar
heat as well as to take advantage of the heat generated within the
building itself. Communities will take on a new look as new buildings turn their windowed faces to the south. More houses and commercial buildings will be earth-bermed, and more will be designed to
function specifically as solar collectors. Such buildings, in which
elegant design will be combined with proper site orientation and
proper glazing, will result in structures whose heating requirements
can be met either with on-site sources (such as wood or propane
stove) or with small amounts of centrally supplied power. The actual
amount of externally supplied energy required for heating or cooling
will be so much smaller than what it is at present, that the exact
kind of energy used will not matter much. And because such a small
amount of energy will be required, it will not be economical to invest in expensive heating and cooling equipment. For example, the
maximum amount of heat required to warm a carefully designed and
constructed 2000-ft^2 residence in Knoxville, Tennessee is equivalent
to the amount of heat from four hair dryers.

ENERGY EFFICIENCY AND RESILIENCE

Are there disadvantages to diminishing the amount of energy that
leaks from our sieve-like buildings? According to some observers,
it *pays* to waste energy, because it would be relatively painless to
quickly tighten our energy belt in the event of a shortfall. Such
reasoning would be analogous to the argument that we should burn

dollar bills at a regular rate during our lives, so that if our income were suddenly cut, we could quickly adjust to the scarcity by no longer burning them! It would make much more sense to tighten our "leaks" now and invest those saved energy dollars by filling the strategic petroleum reserve, for example, or by retrofitting more buildings, or by purchasing high-mileage cars. Then, should an emergency arise, our daily energy demand would be lower, and a given amount of reserve would last longer.

ENERGY SAVINGS AND CONSUMER COSTS

A second and oft-expressed concern about conservation reasons that when people use less energy, the unit price of energy rises, and people end up paying more for the same amount of energy. It is very important to work through this hypothesis carefully, not because it is true (it is not) but because it is a complicated issue.

Let us consider a special case in which this hypothesis could be true--a utility dominated by fixed costs (i.e., a situation in which the total cost of business is not very sensitive to the amount of energy delivered). If demand *drops* for that utility, then revenue will drop, and the only way for the company to stay in business is to raise unit price. This situation takes a rare set of circumstances: a decrease in demand *plus* a situation in which utility costs are not sensitive to output. Short-term reductions in demand can occur, just as annual drops in demand occur in spring and fall. An energy utility's response should be to schedule maintenance for anticipated slack periods and in the case of unanticipated reductions, to turn off the most expensive sources of supply. In the overwhelming majority of situations, however, beyond short transients, energy demand does *not* decrease. Even the most "optimistic" projections of energy futures indicate level or increasing demand, especially for electricity. Hence, if demand reductions occur when normal demand is low, the consumer will likely have to endure short-term increases in unit charges; if reductions occur when normal demand is *high,* the consumer could very plausibly expect a reduction in unit charges.

Far more important than transient effects are the long-term impacts of conservation. During the heyday of inexpensive energy, when the marginal cost (e.g., from the newest power plant) was *lower* than the average cost from the system, it *was true* that "the more we used, the cheaper it got." But that situation no longer exists; the reverse is now true, and every new energy plant or source that is added to the system *raises* the average price. We are in an era that dictates "the more we use, the *more* it costs." This means that the effect of conservation (making demand growth smaller than it would otherwise be) is to hold down the rate of increase in energy price.

Similarly, the energy cost at times of peak demand is always higher than at other times. There are, however, ample opportunities to cut demand at peak (especially at the *daily* peaks) or to defer consumption to off-peak periods. Again, the net effect is to save money and other resources. Ingenuity can be put to work in the construction and operation of buildings so that their energy demands are both small and relatively constant. To the extent that this occurs, we require fewer pumped storage lakes, expensive peaking generators, and the like. Users quickly respond to price signals. It is difficult to imagine, for example, what the effect on the peak load of telephone service would be if time-of-day and day-of-week pricing were not used. The same can happen in the energy sector. If the unit price of energy were set at its marginal cost, then people would gain the incentive to adjust to the situation not only by changing their behavior (as can be illustrated by patterns in telephone use) but also by investing in ways to make more use of off-peak energy availability. Without such market signals, it should surprise no one that energy exhortations to cut peak demand are fruitless.

OPPORTUNITIES THROUGH RESEARCH

While it is true that "energy is always conserved" (first law of thermodynamics), it becomes less valuable after each instance of use (second law). For example, when natural gas is used to heat water, *no* calories are lost, but the inherently very useful calories stored in the gas are converted to hot water and waste heat. In their new form, these calories are far less "available" to do work. Therefore, while energy per se is never lost, its use always results in degrading its *availability* to do work. As energy gets more expensive, the technologist devises ways that utilize smaller amounts of high-availability energy for a given service rendered. He must also take precautions to match the quality of the energy source to the quality of energy needed to perform a given task, so that minimum availability is lost.

Another natural limit to energy efficiency relates the maximum theoretical efficiency of any cyclical process to the combustion temperature T_H and the coolant (discharge) temperature T_C. The *maximum* theoretical efficiency E is:

$$E = 1 - \left(\frac{T_C}{T_H}\right)$$

where temperature is measured in the Kelvin scale ($32^\circ F = 0^\circ C = 273K$). Clearly, unless the combustion temperature is extremely high and/or the discharge temperature extremely low, efficiencies are relatively small. It is possible to use fossil fuels to generate combustion temperatures that result in very high efficiencies (fossil fuels, with air combustion, can burn as hot as $3000^\circ F$), but at such high temperatures and (steam) pressures, there is no material that can readily be used to contain the steam (above about $1300^\circ F$). Such limitations in power plant efficiency can be overcome with the use of a combined cycle, wherein a "topping" cycle consisting of a gas turbine, for example, that is followed by a steam cycle is used. But even by combining such methods, one is still constrained to maximum theoretical efficiencies of less than 70%, with 30 to 40% efficiency actually attained for modern electricity generation plants.

Heat pumps, air conditioners, refrigerators, and freezers are subject to the same laws of thermodynamics. It can be shown that while most heat pumps and air conditioners provide about twice as much heating or cooling energy as input energy (by extracting heat

from outside air, or discharging heat into the outside air), they
could *theoretically* supply about 20 times as much. Hence, they have
an actual efficiency of only about 10% of their theoretical maximum
efficiency. Equipment with considerably improved energy efficiency
could be manufactured, but each improvement in efficiency requires
more investment, and therefore the product will be more expensive.

Such a large fraction of energy end uses actually requires only
relatively low temperatures, it is evident that thermodynamics
matching between energy sources and energy uses holds much promise.
In other words, much better thermodynamic utilization can be obtained
through a variety of techniques, such as total energy systems, cogeneration, and low temperature storage. This notion is discussed
in more detail below.

Other limitations, imposed by natural law, to the efficiency of
energy-consuming devices are set by the properties, such as heat
conductivity, friction, and high-temperature strength, of the different materials used. For example;

1. The efficiency of a refrigerator is strongly dependent on the
   properties of the heat exchanger or radiator which transfers
   heat into the outside air.
2. The finite resistance to heat transfer through insulating materials limits the conservation design for holding heat in--or
   out--in appliances, homes, and industrial processes.

Thus, one of the most fundamental challenges to improving energy
productivity lies in the improvement of the various properties of
existing materials and in the development of new materials and processes with special characteristics. As our scientific understanding of the properties of materials grow, we have been able to make
marked progress in technology and in decreasing the amount of energy
required to provide a given amenity. For example:

1. The progression from vacuum tubes to transistors and thence to
   large-scale, integrated circuits has decreased the amount of
   energy required for many functions in communications and computers more than a millionfold in two decades.

2. The efficiency of electricity conversion into lighting has increased from less than 5% (incandescent) to more than 30% (alkali-
halide) since 1950.
3. The energy required to produce low-density polyethylene has been
halved, thanks to a new process.

In all of these instances, the notion of saving energy has been
an important, but not necessarily dominant, factor in spurring the
development of these processes. Indeed, many similar energy savings
processes were advanced while energy prices were falling. A more
important factor in such research has been the drive to improve
overall quality, cost, and reliability of each product. Nevertheless, close consideration of virtually every major energy-consuming
activity reveals major opportunities for the improvement of energy
efficiency through technological advances. Of course, this generally requires increased capital cost and the successful application
of human ingenuity.

As nations industrialize and urbanize, their citizens tend to
accept conditions of increasing socioeconomic complexity in exchange
for material benefits. The United States is a case in point. The
American consumer generally lives in an environment that is sustained
by advanced technology, largely because the increasing scale of
operations (e.g., farming, communications, power production, transportation) tends to decrease unit production costs. There are, of
course, undesirable but "external" side effects, such as workers
having to commute greater distances (at their own expense). In
other words, increased interdependence between economic units can
result in larger economic savings, albeit with lessened degrees of
freedom.

An analogy can be drawn in the case of energy consumption. The
amount of purchased energy required to provide space conditioning
in a house can be considerably decreased by installing a complex
automatic electronic sensing and control system. However, there a
are three conditions that tend to govern the extent to which these
kinds of conservation tactics are attractive:

1. *Energy Price*. As energy gets relatively more expensive, com-
    plexity and interdependence, which encourage more efficient en-
    ergy use, become more tolerable.
2. *Reliability*. We tend to accept complexity and interdependence
    more readily when the system can be counted on to operate with-
    out unwanted interruption. An apartment owner is not interested
    in being heated by "waste heat" from a power plant that cannot
    be counted on to operate steadily throughout the heating season.
    Improvement in reliability can be expected to continue to re-
    sult from investments in science and technology; however, con-
    sumers are increasingly suspicious of complexities and new high
    technology.
3. *Proximity and scale*. Because of the past history of energy
    prices, economies of scale in industry, and land use patterns
    of residential and industrial development, there are few major
    opportunities for energy savings through increased interdepen-
    dence in our *existing* system. Present sources of low-grade and
    waste heat sufficient to heat and cool buildings are generally
    sited too far from existing buildings to justify transport costs
    of the heat. The extremely large size of power plants causes so
    much waste heat to be generated that markets within plausible
    range are oversaturated. As new capital investments are made,
    much improvement can be expected from energy-conscious siting
    and sizing of energy production and other industrial plants,
    which can result in considerably expanded opportunities for high
    efficiency "cascading" of energy use.

ECONOMICS

Consumers seek to maximize their welfare by spending their income
in ways that provide desired goods and services. Producers, in
turn, seek ways to provide those goods and services in the most attractive (and usually, least costly) fashion, in order to beat their
competitors. As real energy prices rise, both producers and consumers attempt to readjust to the new situation. Over a period of

time that is comparable to capital stock turnover, the amount of
energy required to provide a given amenity can be altered considerably. Because capital investments last a decade or more, the pro-
ducer/consumer must make some assumptions about energy prices over
the (future) lifetime of his investment--an uncertain undertaking,
at best. It would be unwise to presume, as we have in the past,
that future energy prices will be *lower* than what they are at present. Similarly, it is probably unwise to presume that energy prices
are going to be an order of magnitude *higher* than what they are at
present. The challenge to research, development, and demonstration
(RD&D) is to both lower the uncertainty about future energy costs
and provide options to allow much higher energy efficiency of use
for minimum increases in capital cost, thereby desensitizing capital investments to the inroads of higher energy prices.

At the present time, the energy market price and the price of
energy-intensive activities such as transportation are so distorted
by public policies that it is difficult to discern real total costs.
It is even more difficult to forecast the real cost that can be
used in trade-off economic studies. Numerous analyses indicate that
the investment required to *save* a unit of energy (e.g., through the
uses of insulation, improved process controls, high-performance automobiles) while still providing the amenity is considerably less than
the investment required to *add* a unit of energy (e.g., through power
plants, coal gasification plants, solar collectors).

A RD&D AGENDA FOR CONSERVATION

There is every reason to believe that further RD&D in energy efficiency will continue to create additional cost-effective options.
The *goals* of a RD&D program in conservation should be to

1. Provide new options for technological innovation, with federal
   emphasis on those options that cannot logically be expected
   from the private sector

2. Develop new and modify existing institutional mechanisms (e.g.,
   regulatory measures) to facilitate, rather than retard, the

development and adaptation of energy and other resource-conserv-
ing procedures and technologies

3. Ensure that such new options are either benign, from an environ-
    mental and health aspect, or are no more injurious to the eco-
    system than the alternative of using more energy

4. Slow demand growth in a way that can lead to a long-term, sus-
    tainable energy system, while continuing to help provide de-
    sired amenities

Conservation RD&D *strategies* to achieve these goals in the
buildings sector should include the following objectives:

1. Develop an extensive series of basic research activities
aimed at improving our fundamental understanding of the physical,
chemical, and biological properties of materials.

2. Develop increased opportunities for energy transformation
and conversion at point of use. Of course, this option has limited
applicability in a nuclear economy in which conversion can only be
done at large central plants. The tendency toward electrification by
using massive, single-purpose central station plants precludes most
opportunities to make productive use of the "waste" heat, which
presently amounts to about two-thirds of input energy. Even with
very advanced technology, about half of the potentially valuable
heat energy would be thrown away at a power plant. Most power plants
are so large and are sited so remotely from cities and industry that
there is little occasion to use the reject heat. If energy conversion were done instead in smaller units that are located close to
the points of use, there would exist many more opportunities for
the productive use of more of the total energy resource. To this
end, there are three general approaches: greater emphasis on producing gases and liquids and lesser emphasis on large central station generation of electricity; decentralized production of electricity (e.g., smaller size fossil-fueled plants, direct solar fuel
cells); and siting of moderate-to-large size power plants close
enough (less than 50 mi) to industrialized and urbanized areas, so
that "waste" hot water can be economically piped to various consumers.

There are relatively few unique needs for electricity, a rather expensive form of energy. A gas-fired heat pump can heat and cool a building and simultaneously provide hot water and heat for cooking. Gas can be obtained from natural sources or synthesized from coal, biomass, and biological wastes. However, in the case of gas synthesized from coal, we are not certain that coal-SNG-heat pumps are more efficient than coal-electricity-heat pumps. These trade-off issues are complex and merit considerable study. One central challenge in developing this kind of diverse and dispersed energy system is to hold capital costs within economic bounds. *Therefore, a central technological challenge is to learn how to do things efficiently on a small scale--a considerable departure from our past focus on achieving economy by going to larger scale.*

3. Identify possible and plausible future spurs to energy demand. Much of the growth in energy demand since 1950 can be attributed to the advent of new, energy-intensive consumer goods and services, coupled with rising incomes and falling energy prices. Since 1950, many energy-intensive products have been introduced to consumers, and many are already well on their way toward market saturation. These include home freezers, television, and air conditioning. Further increases in automobile ownership and travel are limited, simply because we now have almost one car for every driver and now spend, on the average, almost 1 hr/day in a car. Illumination, refrigeration, water heating, and space heating are now universal in the United States, and the markets for these are now mostly of the replacement type.

Some observers caution that these facts should not lead us to assume a consequent slowdown in the growth of energy demand. Rather, they argue, new, energy-intensive items that will keep energy (especially electricity) demand growth high will continue to emerge. One such "item" is "outdoor space conditioning" and enclosed cities (although the latter could actually be less energy-intensive than the alternative of separately space conditioning all the individual buildings). To be sure, there could be such developments, however

strongly they be coupled to energy price behavior. It follows, then,
that more thought must be given toward identifying the consumer's
unfulfilled but inherently energy-intensive desires and toward fac-
toring these desires into planning models that project future demand. In addition, evolving social values could lead some segments
of our society to seek *less* energy-intensive life styles.

    4. Develop energy conservation technology export markets. Our
international relations with other industrialized countries, oil-
exporting countries, and developing nations are strongly affected
by our energy conservation RD&D policies and by our progress in
utilizing these policies. Because the United States' call on world
supplies of energy are influenced by our domestic developments, other
oil-importing nations watch our policies with great interest. Since
1968, the United States has maintained a zero domestic energy supply
growth by turning to imports, while at the same time doing little to
slack its increasing energy thirst. Energy-importing countries,
however, provide extremely important market opportunities for U.S.
conservation technology, and the fact is that our efforts to assist
lesser developed countries in their economic development could
benefit by the implementation of energy-efficient innovations that
are adaptable to less technically sophisticated societies.

MAKING THE TRANSITION

There are three distinct approaches to altering the energy use characteristics of our buildings. Each approach has a different time-
constant, although these constants overlap. The total effect of
using each approach is a major energy saving. Substantial savings
appear early in the three cycles, but the full impact of savings requires decades, the time for turnover of our stock of buildings.

    The first approach, a no-to-low capital cost-charge effort in
building operation and maintenance is already well underway, although it has a long way to go. The program ranges from the installation of a very inexpensive flow restrictor in a shower head to
the cleaning of heating, ventilation, and air conditioning (HVAC)

equipment and the installation of controls that will automatically
turn heating/cooling systems on and off as desired. In the case of
commercial buildings that are actually used approximately 50 hr/week,
initial low-cost investments in the HVAC system might entail checking and balancing the system and installing a simple control system
that conditions the building space before the tenants arrive and
shuts it down after business hours (such systems can be equipped
with simple or complex override mechanisms that adapt to individual
work patterns). On the whole, this does not represent a large capital investment and yet substantially improves the energy productivity of the building. Similarly, since many residences are occupied
for no more than 14 hr/day, the use of automated thermostat controls can provide major energy savings, and at *no* loss in amenity
level for the occupants. These kinds of actions should make good
sense to virtually every utilities-paying owner or occupant, regardless of economic condition. They provide quick, substantial savings
for very little investment and can be implemented rapidly.

The second approach now gaining momentum and subject to a growing number of state and federal subsidies and support is a major
capital retrofit investment effort. In a commercial building, this
could take the form of thermal insulation, a new HVAC system with
(thermal) sensing and monitoring capability, and, perhaps, some external treatments, such as sunshading. Older buildings may require
extensive checks and modifications to block infiltration. In a
home, the addition of storm windows, replacement of a boiler or furnace, adjustments to block thermal bypasses and to remedy other infiltration problems, and, perhaps, the addition of some attic insulation may be indicated. These choices should be made on the
basis of an inspection by a trained energy auditor or other person
familiar with the technical and economic choices.

In many locations, retrofitting with passive solar features,
such as Trombe walls and solar greenhouses, are cost-effective.
People with homes oriented to the north may consider adding a south-
facing family room to replace a seldom-used north-facing living room.

The addition of vestibule entries is likelty to become more common.
Such retrofit measures, when combined with normal "standard" options,
such as appropriate caulking and weather stripping, can achieve
space conditioning efficiency improvements of 50%, under most circumstances. As air-to-air heat exchangers for residential use enter
the market, even greater efficiencies may be safely achieved, because the buildings can then be made airtight.

   For the intermediate future, the most substantial gains in energy efficiency in buildings can be achieved through a retrofit
effort, since more than two-thirds of the existing building stock is
expected to be still in use in the year 2000. Other than improvements in the transportation sector (especially in automobiles), this
opportunity represents the nation's fastest and most cost-effective
way to deal with the energy problem in this time frame. It is particularly urgent that we work on buildings, such as those in much of
the northeastern United States, that use liquid fuels. For these
buildings, two measures can be employed separately or together:
switching to a different energy source, such as electricity, and/or
insulating and otherwise lowering the energy requirements of the
structure.

   In terms of dramatic efficiency improvements, the greatest
unit contributions and the most important long-term gain will be in
new construction, the third approach. The market demand for buildings with high-energy efficiency, bolstered by federal regulations,
such as the Building Energy Performance Standards, will slowly but
surely take the edge off of energy demand growth in the buildings
sector. Despite a growth in number and total volume of buildings,
the amount of purchased energy that will be required to service
them 30 years hence need be no more, and can perhaps, be less than
the amount that is required today.

   Interestingly enough, the design and construction of new buildings of vastly improved energy efficiency will not be substantially
more capital-intensive than are present buildings. This is because
the increased investments in insulation, triple-glazed windows, and

the like are partly offset by lower investment requirements for
furnaces, air conditioners, and so on. Perhaps more significant
is the fact that the *alternative* of adding to our energy supply
would constitute an even greater investment, and this is without
considering such additional external costs as environmental degrada-
tion.

The central policy questions with the most far-reaching implications for the future of our country are the relative priority assigned to energy/buildings questions and the rate at which the transition from a less efficient to a more efficient economy is undertaken. If we intend to radically alter energy demand in the buildings sector by the 1990s, a circumstance that is technically possible
and highly desirable in terms of economic and national security interests, a combination of rising energy prices, strong regulation,
and increased direct federal intervention (weatherization and similar retrofit assistance, for example) will be required. This will
undoubtedly mean a change in federal-state relations and will affect
private market mechanisms in more ways than have traditionally occurred, except in times of national crisis. Such a program could
be achieved cost-effectively, would vastly improve the national
security posture of the country, and could more quickly be implemented than meeting energy demands with supply-side approaches.
However, barring another set of severe shocks to the system, such
as the 1973 oil embargo, it is not likely that such a program will
be put into affect.

A more likely policy is that of steady but slower movement
that is sustained by the pricing of fuel to reflect true replacement costs and increasing regulatory actions. Since the principal
objectives to be served by aggregating the fuel savings are national,
it is likely that the federal government will continue to be the
prime regulator. Appliance standards, new building standards, and
some type of retrofit requirements can be expected. Accompanying
federal subsidies will sweeten the prospect, as continuing concerns
over the equity aspects of rising prices are likely to be resolved

through substantial federal assistance for the improvement of the
quality of housing, the effect of which will make the low-income
consumer more resilient to energy shortages over the long run.
Furthermore, federal actions that influence rate and pricing policies of utilities will continue, and efforts to develop a rational
national fuel use policy are likely to show results.

Federal subsidies in the form of research on basic materials
and on heat transfer problems and independent research which focus
on other questions and are responsible for spin-off results will
identify entirely new methods for heating and cooling. It is conceiv-
able that before the end of the century, substantial price and performance breakthroughs in photovoltaic technology will radically
alter the nature of electric utilities and allow for the construction of neighborhood or community grids. The interconnection of
grids nationally, combining power generated by thermal capacity and
renewables, may also occur.

However rapidly we move, crafting the national energy policy
will call for new levels of flexibility that will serve both the national need to conserve energy rapidly and the varying needs of
states, localities, and individuals. Policymaking of this type is
new for our country, which has preferred to set broad guidelines
and allow wide variation in the application of those guidelines.
Our capacity to make the necessary choices wisely will reflect our
thoughtfulness in our approach, the willingness of each level of
government to understand all points of view, and the degree to which
the policymakers recognize different levels of need and concern.
Our traditional American pluralistic approach to solving problems--
trying out a number of paths and relying on market and social mechanisms to define many solutions--is on our side in this effort. A
good-faith commitment to experiment, to test, and then to choose
policies that meet as many of our national objectives as possible
is needed. I am confident that our system of government can rise
to this challenge.

NOTES AND REFERENCES

1. Office of Technology Assessment, *Residential Energy Conservation*, U.S. Congress, July 1979.

2. John H. Gibbons, Long term research opportunities, in *Energy Conservation and Public Policy*, John Sawhill, ed., Prentice-Hall, Englewood Cliffs, N. J., 1979.

3. *Alternative Energy Demand Futures to 2010*, Report of the Demand and Conservation Panel to the Committee on Nuclear and Alternative Energy Systems, National Research Council, 1979.

4. U.S. energy demand: Some low energy futures, *Science*, vol. 200, April 14, 1978, pp. 142-152.

List of Figures

Fig.   Page

1-1   Buildings energy use by fuel type, 1950-1978   6
1-2   Energy use in residential and commercial buildings,
      1950-1978   6
1-3   Trends in real fuel prices to the residential sector,
      1950-1978   7
1-4   1975 residential energy use .per household   10
1-5   1975 commercial energy use per square foot   11
1-6   Predicted energy savings for several thermostat settings   13
1-7   Effects of energy conservation practices and measures on
      annual energy use in an office building constructed
      during the 1960s   16
1-8   Annual space heating load for a new Kansas City single-
      family home versus additional initial cost   18
1-9   Relationships between annual heating and cooling loads
      for a new Kansas City single-family home   18
1-10 Annual energy use versus initial cost for a typical gas
      furnace   19
1-11 Annual energy use versus initial cost for a typical
      refrigerator   20
1-12 Engineering analysis of HVAC systems   21
1-13 Projected fuel prices with and without NEA, 1970-2000   28
1-14 Baseline projection of residential energy use, 1970-
      2000   29
1-15 Baseline projection of commercial energy use, 1970-2000   29
1-16 Projected savings in the buildings sector due to NEA
      and RD&D, 1975-2000   35
1-17 Projected residential energy savings due to NEA and RD&D,
      1975-2000   35

433

Fig.   Page

2-1   Energy applications   42
2-2   Effect of illuminance and task difficulty on visual
       performance, age 25, task demand level 45.0   48
2-3   Effect of illuminance and task difficulty on visual
       performance, age 25, task demand level 75.0   48
2-4   Effect of illuminance and task difficulty on visual
       performance, age 65, task demand level 45.0   49
2-5   Effect of illuminance and task difficulty on visual
       performance, age 65, task demand level 75.0   49
2-6   Clerical workers' reactions to lighting levels   50
2-7   Veiling reflection   51
2-8   Energy savings achieved by turning off lights during
       unoccupied hours   54
2-9   Office using ellipsoidal reflector lamps   56
2-10   Ellipsoidal reflector lamp   56
2-11   Light efficiencies   59
2-12   High-pressure sodium lighting   60
2-13   Lamp depreciation characteristics   61
2-14   Lamp characteristics   63
2-15   Lighting control system   67
2-16   Recessed fixtures   71
2-17   Regressed fixtures   73
2-18   Systematic lighting maintenance   76
2-19   Luminaire depreciation   77

3-1   Residential energy use in the United States   90
3-2   Air conditioning EER of heat pumps and central air
       conditioning units shipped in 1977   106
3-3   Energy flows in a 40-gal water heater   110
3-4   Energy flows in a 50-gal electric water heater   110
3-5   Heat pump energy savings   116
3-6   Refrigerator heat gains   118
3-7   Refrigerator electricity consumption   119
3-8   Energy use versus retail price for a top freezer-
       refrigerator   125

4-1   Schematic of a solar heating and cooling system   147
4-2   Schematic of a domestic hot water system   148
4-3   Air collector panel   151
4-4   Flat-plate liquid collector panel   153
4-5   Trickle-down collector   153
4-6   Evacuated tube collector   154
4-7   Yearly performance based on collector orientation   155
4-8   Storage subsystems   157
4-9   Heating directly from collectors of an all-air system   159
4-10   Space heating from storage   161
4-11   Storing heat   162
4-12   Heating from the collector   163
4-13   Heating from storage   165

Fig. Page

4-14 Heat storage mode   166
4-15 Typical thermosiphon solar water collector   167
4-16 Roof-mounted pumped solar water heating system   168
4-17 Dual liquid/solar hot water heater   170
4-18 Absorption air conditioner   173
4-19 Solar/Rankine-powered air conditioner   174
4-20 Air-to-air heat pump with solar assist   176
4-21 Water-to-air heat pump with solar assist   177
4-22 Fraction of annual heating load furnished by a solar
      heating system   180
4-23 Average daily solar radiation (Btu/ft^2) month of
      January   182

5-1   Meigs' building plan detail   198
5-2   Pueblo cliff dwelling   200
5-3   Energy consumption patterns   202
5-4   Energy and form   206
5-5   Marsh passive solar house--Jamaica   208
5-6   Marsh solar dwelling--Nova Scotia   210
5-7   Leela design house   219
5-8   Kelbaugh house   220
5-9   Calthorpe cabin   222
5-10 Rust house   224
5-11 Georgia Power Company Building   226
5-12 Benedictine Publications office and warehouse   228
5-13 Aspen Airport   229
5-14 Faneuil Hall Marketplace   231

6-1   Building heat flow   240
6-2   Hypothetical IEA building   245
6-3   Total monthly energy consumption for heating and cooling
      in building (with internal loads)   248
6-4   Annual peak heating demand in building (with internal
      loads)   249
6-5   Annual peak cooling demand in building (with internal
      loads)   249
6-6   Net thermal load (heating-cooling) in building (with
      internal loads)   250
6-7   Hourly load profiles of building for cyclic summer day
      (clear sky) (heating-cooling)   251
6-8   Total monthly energy consumption for heating and cooling
      in building   253
6-9   Annual peak heating demand in building (with internal
      loads)   254
6-10 Annual peak cooling demand in building (with internal
      loads)   254
6-11 Hourly cooling loads from internal heat gains   260
6-12 Total monthly energy consumption for heating and cooling
      in building   261

Fig.   Page

6-13 Annual peak heating demand in building (with internal
      loads)   261
6-14 Annual peak cooling demand in building (with internal
      loads)   262

7-1   Building components functional relationships   272
7-2   Chimney effect   277
7-3   Infiltration   278
7-4   Landscaping and orientation   279
7-5   Internal shading   281
7-6   External shading   281
7-7   Horizontal overhang   282
7-8   Louvered windows   283
7-9   Sawtooth walls   283
7-10 Composite of schematic site plan   286
7-11 Climate zones of the United States   295
7-12 Icehouse roof   300
7-13 Skylids   301
7-14 Sky Therm roof   302
7-15 Total annual primary energy use per unit for charac-
      teristic and improved residences   307
7-16 Total annual primary energy use per square foot of
      floor area for characteristic and improved residences   308
7-17 Minimum energy dwelling   311
7-18 The comfort control system   313
7-19 Diagrammatic section of LER house   316
7-20 Cross-section view   317
7-21 ACES house   323
7-22 Space heating schematic   325
7-23 Comparison of ACES and control house week beginning
      1/9/78   328

8-1   Single duct system   341
8-2   Terminal reheat system   342
8-3   Multizone system   343
8-4   Dual duct, low-velocity system   344
8-5   Variable volume system   346
8-6   Induction unit system   347
8-7   Retrofit insulation installation to on-grade floors   352
8-8   Solar screening devices   358
8-9   Mediterranean shutter design   359
8-10 Illumination versus power density   364
8-11 Retrofit for recycling exhaust air   368
8-12 Retrofit of kitchen exhaust hoods   369
8-13 Enthalpy wheel--summer cooling   371
8-14 Typical central control system   380
8-15 Buildings central control system--programs and
      interface   381

9-1   Illustration of energy benefits   398
9-2   Benefit/risk trade-off criteria   400

List of Tables

Table   Page

1-1 The residential sector, 1950-1978   3

1-2 Approximate differences in residential space heating
energy use by housing type   4

1-3 The commercial sector; 1950-1978   5

1-4 Residential energy use by fuel/end use   8

1-5 Commercial energy use by fuel/end use, 1975   8

1-6 Residential and commercial energy use by federal region,
1975   9

1-7 Effects of natural ventilation on residential air conditioning requirements   14

1-8 Inputs used in projections of residential and commercial
energy use to year 2000   27

1-9 Alternative residential and commercial energy projections: Energy use   32

1-10 Alternative residential and commercial energy projections: Direct economic effects   33

2-1 Lighting applications   43

2-2 Lighting   44

2-3 Cost analysis based on time required to perform work when
lighting is reduced from 150 to 50 fc   45

2-4 Productivity and cost analysis on government check
reading study for three lighting levels   46

2-5 Illuminance categories and values for generic types,
of activities in interiors   53

2-6 Conversion to Maxi-Miser for 500 troffers   58

Table		Page
2-7	Lighting cost comparison	74
2-8	High-performance lighting systems and other fixtures	78
2-9	Comparative costs of skylights, fluorescent lights, and the two in combination	81
2-10	Efficiency comparison of skylights and electric lighting	82
3-1	Annual energy requirements of electric household appliances	92
3-2	Estimate of average annual energy consumption of electric appliances	94
3-3	Estimate of average annual energy consumption of gas appliances	96
3-4	Estimate of appliance average annual operating costs	98
3-5	Eight popular heating systems in existing single-family homes	100
3-6	Heating systems energy savings with various design changes	102
3-7	Estimates of the SPF for three advanced gas heating systems	105
3-8	Room air conditioner design options to improve efficiency	108
3-9	Energy and economic effects of increased jacket insulation thickness	112
3-10	Energy and economic effects of reduced jacket insulation thermal conductivity	113
3-11	Expected energy savings and economics for heat pump water heaters	115
3-12	Evaluation of energy saving options for 16-ft^3 automatic defrost, combination refrigerator-freezer	122
3-13	Energy savings achievable for representative refrigerator models	124
3-14	Percentage of savings with microwave versus conventional electric range	127
3-15	Summary of technological improvements in ovens and ranges	128
3-16	Likely classes and tentative determinations of maximum technologically feasible energy efficiency levels	134
4-1	Cost of heat from passive solar heating installations	145
4-2	Types of collectors	150
4-3	Rules of thumb for sizing	180
4-4	System costs according to function and manufacture	185
4-5	Nature, frequency, and significance of technical malfunctions	187
4-6	Numbers and types of projects	188
4-7	Apparent solar fraction	190
4-8	Measured auxiliary/predicted auxiliary use with solar	192
4-9	Profile of measured losses	194

Table   Page

6-1   Characteristics of load programs   241
6-2   Hypothetical building characteristics   244
6-3   Building variations   246
6-4   Programs analyzed   246
6-5   Comparison of annual heating and cooling requirements   248
6-6   Amount of heat transmitted through glass due to tem-
       perature differences on cyclic (repeating) days   255
6-7   Amount of net space solar energy gain through glass
       on cyclic summer day   256
6-8   Annual heating and cooling loads   259
6-9   Annual loads for building variation C   260
6-10   Annual loads for building variation B   262
6-11   Program methodologies   263
6-12   Amount of heat transmitted through glass due to tem-
       perature differences on winter repeating day   264
6-13   Relationship between heat transmission through glass
       due to temperature differences and the convective-
       radiative split of the indoor film coefficient   265
6-14   Amount of net space solar energy gains through glass
       on cyclic summer day   266

7-1   Residential energy use in Midwestern United States   272
7-2   Breakdown of heating and cooling loads for character-
       istic house   273
7-3   Buildings configurations options: Physical char-
       acteristics   288
7-4   Buildings configuration options: Energy consumption
       characteristics   289
7-5   Types of thermal building insulation   291
7-6   Insulation value of common materials   292
7-7   Insulation guidelines for homes heated with oil, gas,
       or heat pumps   294
7-8   Insulation guidelines for homes heated with electric
       resistance heat   294
7-9   R-values chart   295
7-10   Summary of new fenestration technologies   298
7-11   Characteristic single-family residences annual energy
       data   305
7-12   Modifications to single-family buildings   306
7-13   Calculated building envelope performance   318
7-14   Estimated annual fuel usage for space heating   319
7-15   Construction cost estimate   322

8-1   Commercial energy use, 1950-1975   332
8-2   Commercial building inventory   333
8-3   Commercial inventory by region   334
8-4   Commercial energy use by building type, 1975   335
8-5   Commercial energy use by fuel/end use, 1975   336
8-6   Commercial energy use by end use   336

Table   Page

8-7   Insulation value for heat flow through opaque areas
       of roofs and ceilings   353
8-8   Insulation value for heat flow through opaque exterior
       walls for heated areas   354
8-9   Yearly heat gain per square foot of single- and
       double-glazing   356
8-10   Visual difficulty rating (R) of tasks   362
8-11   GSA-recommended lighting levels   362
8-12   General lighting levels   363
8-13   Ventilation requirements for occupants   366

9-1   Annual GNP and energy consumption growth rates   392
9-2   World energy growth rates: History and projections,
       Series A-E, 1960-1965   394

Index

ACES house, 321-329
Air conditioning, 105
Appliances, 89
  efficiency standards, 30, 133-
    136, 406
  integrated, 132
Arab Oil Embargo, 201
Arkansas house, 317
ASHRAE, 247, 252, 258, 365, 383
Atrium, 302-303

Baer, Steve, 143
Balcomb house, 144
Ballasts, 77-79
Beadwall, 230
Brownell LER house, 315-321
Buchanon, Deborah, 144
Building
  analysis techniques, 383-384
  configuration, 287-289
  envelope, 285-287
  insulation, 290
  management and control, 379-
    381
  performance standards, 31,
    213, 406
  shape, 350

[Building]
  siting, 247, 350
  thermal storage, 247

Calthorpe, Peter, 222
Chicago Navy Pier, 232
Chimney effect, 338
Clothes dryer, 131-132
Clothes washer, 131-132
Coefficient of utilization, 62
Commercial buildings, 2, 333-337
  building stock, 333
  energy use, 8, 331
  floorspace, 4
Computer analysis system, 237
Conservation
  barriers to, 22, 395-396
  residential, 2
  R&D agenda, 423-426
Contrast rendition, 50-52

Daylighting, 79-82
Demand controllers, 69
Dimmers, 69
DOE-1, 247, 251, 252, 255
Dubin, Fred, 348, 351, 355, 357,
  360, 376

EBTR, 407
ECUBE-75, 248, 250
Electric igniter, 104
energy
  availability, 419
  consciousness, 204
  cost, 413
  efficiency, 419
  prices, 412
  programs, 407
  use, 5
Energy Conservation and Production Act, 212
Enthalpy wheel, 370
Equivalent spherical illumination, 51
ESA, 248, 251

Faneuil Hall marketplace, 230
FEMP, 407
Flue dampers, 100
Freezers, 117-125
Fuel prices, 7

Gas furnaces, 19
Government conservation programs,
  30
Greenhouse, 302-303

Hay, Harold, 222, 300
Heat pipe, 370
Heat pump, 105, 370, 377
  solar assisted, 175-178
Henderson, John K., 365
Households, 2
HVAC, 20, 91-96, 304, 338-339,
  367, 372-376
  central plant, 376-378
  multizone system, 340
  operations and maintenance,
    378-379
  single zone system, 340
  variable air volume system,
    345

Illuminance, 47
Illumination
  non-uniform, 64-65

Insulation, 353

Jamaica house, 210
JULOTTA, 252, 255

Kelbaugh house, 220

Lamps, 55
  lumen depreciation, 72
  maintenance, 72-77
Leela design, 218
Life cycle cost, 24, 414
Lighting, 284-285, 361-364
  age, 47-50
  controls, 65-69
  cost benefit, 45
  fluorescent, 57-62
  high intensity discharge, 57
  high pressure sodium, 63
  incandescent, 57
  management, 42
  metal halide, 62
  sources, 55-64
  systems, 2
Load modeling, 239-244
Luminaires, 70-72

Marsh, Mike, 207
Meigs, General Montgomery, 197
Michel, Jacques, 143
Minimum energy dwelling, 309-315

National Energy Act, 26

Office of Technology Assessment,
  413
Oil burner
  retention head, 101
  variable firing range, 101
ovens, 126-130

Pension Building, 197, 203, 234
Potter, David, 385
Power factor correction, 69

Ranges, 126-130
Refrigerators, 19, 117-125
Residential
    energy use, 8
    tax credit, 404
Residential Conservation Service,
    404
Roof design, 297-301
Rust, Thomas, 223
R Value, 290

Sage, Andrew, 402
Sawyer, Stephen, 184
Schools and Hospitals Program,
    405
SCOUT, 247
Seasonal performance factor, 91
Shading, 280-284
Shutters, 357
Skylids, 230, 301
Sky Therm, 222, 300
Solar, 158
    active systems, 146-149
    air systems, 158-160
    cooling systems, 171-174
    direct gain, 142
    hybrid passage, 144
    indirect gain, 142-143
    liquid systems, 160-164
    passive, 140-142, 205, 217

[Solar]
    Radiation, 275
    System sizing, 178-184
    water heating, 164, 171
Solar collector, 146, 149-155
    evacuated tube, 152
    flat plate, 149
    linear focusing, 150
    trickle down, 152
Space heating, 7, 12, 96-105

Thermal storage, 146, 155-158
Trombe, Felix, 143
Trombe wall, 207, 227

Veiling reflections, 50
Ventilation, 14, 276-280, 365-371
Visual comfort, 42, 52

Water heating, 9, 108-109
    heat pump, 114-116
Weatherization Assistance Program,
    404
Wind effects, 276-280
Windows, 295-297, 354-357

ZOMEWORKS, 227, 300